PROGRAMMING AND CUSTOMIZING THE AVR MICROCONTROLLER

PROGRAMMING AND CUSTOMIZING THE AVR MICROCONTROLLER

Dhananjay V. Gadre

McGraw-Hill

New York   San Francisco   Washington, D.C.   Auckland   Bogotá
Caracas   Lisbon   London   Madrid   Mexico City   Milan
Montreal   New Delhi   San Juan   Singapore
Sydney   Tokyo   Toronto

Library of Congress Cataloging-in-Publication Data

Gadre, Dhananjay V.
    Programming and customizing the AVR microcontroller / Dhananjay V. Gadre.
      p. cm.
    ISBN 0-07-134666-X
    1. Programming controllers 2. RISC microprocessors. I. Title.

    TJ223.P76 G33 2000
    629.8'95416—dc21   00-059452

Atmel®, AVR®, and AVRStudio®, among others, are the registered trademarks of Atmel Corporation, 2325 Orchard Parkway, San Jose, California 95131

McGraw-Hill

A Division of The McGraw·Hill Companies

2 3 4 5 6 7 8 9 0   DOC/DOC   0 9 8 7 6 5 4 3 2 1

P/N 134667-8
PART OF
ISBN 0-07-134666-X

The sponsoring editor of this book was Scott Grillo. The editing supervisor was Sally Glover, and the production supervisor was Pamela Pelton. It was set in Times New Roman PS per the TAB4 Design by Deirdre Sheean and Joanne Morbit of McGraw-Hill's Professional Book Group composition unit, Hightstown, New Jersey.

Printed and bound by R. R. Donnelley & Sons Company.

This book is printed on recycled, acid-free paper containing a minimum of 50% recycled, de-inked fiber.

McGraw-Hill books are available at special quantity discounts to use as premiums and sales promotions, or for use in corporate training programs. For more information, please write to the Director of Special Sales, Professional Publishing, McGraw-Hill, Two Penn Plaza, New York, NY 10121-2298. Or contact your local bookstore.

Dedication

To Sangeeta

How many kisses satisfy?
How many are enough, and more?

—Catullus

CONTENTS

List of Figures   *xiii*

List of Tables   *xxi*

Acknowledgments   *xxiii*

Chapter 1 Introduction   *1*

    1.1 Microcontroller, Microcomputer or Microprocessor? *2*
    1.2 Do you need a Microcontroller? *3*
    1.3 Why the Atmel's AVR Microcontroller? *5*
    1.4 Organization of This Book *6*
    1.5 Timing Diagram Conventions *6*

Chapter 2 Microcontrollers   *11*

    2.1 Microcontroller Architecture *14*
    2.2 Choosing a Microcontroller *16*
    2.3 Developing Applications with a Microcontroller *18*

Chapter 3 The AVR RISC Microcontroller Architecture   *21*

    3.1 Introduction *21*
    3.2 AVR Family Architecture *22*
    3.3 The Register File *25*
    3.4 The ALU *26*
    3.5 Memory Access and Instruction Execution *27*
    3.6 I/O Memory *27*
        3.6.1 SREG: Status Register *28*
        3.6.2 SP: Stack Pointer Register *29*
        3.6.3 GIMSK: General Interrupt Mask Register *29*
        3.6.4 GIFR: General Interrupt Flag Register *29*
        3.6.5 MCUCR: MCU General Control Register *30*
        3.6.6 MCUSR: MCU Status Register *30*
        3.6.7 TCCRO: Time/CounterO Control Register *31*
        3.6.8 TCNTO: Time/CounterO Register *31*
        3.6.9 TCCR1A: Timer/Counter1 Control Register A *32*
        3.6.10 TCCR1B: Timer/Counter1 Control Register B *33*
        3.6.11 TCNT1H, TCNT1L: Timer/Counter1 *34*
        3.6.12 OCR1AH, OCR1AL: Timer/Counter1 Output Compare Registers *35*
        3.6.13 OCRIBH, OCR1BL: Timer/Counter1 Output Compare Registers *36*

3.6.14 ICR1H, ICR1L: Timer/Counter1 Output Capture Registers *37*
3.6.15 WDTCR: Watchdog Timer Control Register *37*
3.6.16 EEAR: EEPROM Address Register *37*
3.6.17 EEDR: EEPROM Data Register *38*
3.6.18 EECR: EEPROM Control Register *38*
3.6.19 PORTB: PortB Data Register *39*
3.6.20 DDRB: PortB Data Direction Register *39*
3.6.21 PINB: Input Pins on PortB *39*
3.6.22 PORTD: PortD Data Register *39*
3.6.23 DDRD: PortD Data Direction Register *39*
3.6.24 PIND: Input Pins on PortD *39*
3.6.25 SPI I/O Data Register *39*
3.6.26 SPI Status Register *39*
3.6.27 SPI Control Register *40*
3.6.28 UART I/O Data Register *40*
3.6.29 UART Status Register *40*
3.6.30 UART Control Register *41*
3.6.31 UART Baud Rate Register *42*
3.6.32 ACSR: Analog Comparator Control and Status Register *42*
3.7 The EEPROM *43*
3.8 The I/O Ports *45*
3.9 The SRAM *46*
3.9.1 Interface to External SRAM *47*
3.10 The Timer *47*
3.11 The UART *49*
3.12 The Interrupt Structure *53*
3.13 The Internal Watchdog Timer *55*
3.14 Power-Down Modes of Operation *56*
3.15 Different Types of AVR Controllers *57*

Chapter 4 The AVR Instruction Set   *59*

4.1 Program and Data Addressing Modes *59*
4.1.1 Register Direct (Single Register) *59*
4.1.2 Register Direct (Two Registers) *61*
4.1.3 I/O Direct *61*
4.1.4 Data Direct *61*
4.1.5 Data Indirect *62*
4.1.6 Indirect Program Addressing *62*
4.1.7 Relative Program Addressing *62*
4.2 Arithmetic and Logic Instructions *63*
4.3 Program Control Instructions *67*
4.4 Data Transfer Instructions *72*
4.5 Bit and Bit-test Instructions *76*

Chapter 5 AVR Hardware Design Issues   *81*

5.1 Power Source *81*
5.1.1 Battery Power *82*
5.1.2 Main Operating Supply *83*
5.1.3 Power from Port Signal Lines *84*
5.1.4 Voltage Regulators *85*
5.2 Operating Clock Sources *86*
5.2.1 Using a Crystal Clock IC *86*

5.2.2 Using a Ceramic Resonator 87
5.2.3 Using a Quartz Crystal 88
5.2.4 Using a Quartz Clock Crystal 90
5.2.5 Using Internal RC Clock Oscillator 90
5.3 Reset Circuit 93

Chapter 6 Hardware and Software Interfacing with the AVR   97

6.1 A Beginner's Circuit 97
6.2 Lights and Switches 99
6.3 Stack Operation in AVR Processors 101
6.4 Implementing Combinational Logic 104
6.5 Connecting the AVR to the PC Serial Port 105
6.6 Expanding I/O 110
    6.6.1 I/O Expansion Using Shift Register 110
    6.6.2 IIC Expanders 111
6.7 Interfacing Analog to Digital Converters 112
    6.7.1 AD Conversion Using the On-Chip Comparator 113
    6.7.2 MAX186 117
    6.7.3 MAX186 Data Conversion and Readout 118
    6.7.4 MAX110/MAX111 121
6.8 Interfacing Digital-to-Analog Converters 124
    6.8.1 Using PWM for a DAC 124
    6.8.2 R-2R Ladder DAC 124
    6.8.3 MAX521 DAC 126
    6.8.4 Data Transfer to a MAX521 127
6.9 Interfacing LED Displays 132
    6.9.1 Seven-Segment Displays 132
    6.9.2 Dot Matrix Displays 133
6.10 Interfacing LCD Displays 135
6.11 Driving Relays with AVR 138
6.12 Stepper Motor Interface for the AVR 140
6.13 Interfacing to a Serial EEPROM 141
6.14 Interfacing to a Real Time Clock (RTC) 146
6.15 Accessing a Constants Table 149
6.16 Arbitrary Waveform Generation 150
6.17 A Switch-Case Implementation 150
6.18 Implementing a Finite State Machine 152
6.19 Generating Random Numbers 154

Chapter 7 Communication Links for the AVR Processor   157

7.1 Introduction 157
7.2 RS-232 Link 158
7.3 RS-422/423 Link 160
7.4 RS-485 Link 161
7.5 SPI and MICROWIRE Bus 163
7.6 IIC Bus 164
7.7 PC Parallel Port 166
7.8 ISA Bus 172
7.9 Universal Serial Bus 174
7.10 IrDA Data Link 178
7.11 CAN (Controller Area Network) Bus 182

Chapter 8 AVR System Development Tools 185

8.1 Code Assembler *185*
    8.1.1 AVR Family Assembler *186*
    8.1.2 IAR Assembler *187*
8.2 Code Simulator *187*
    8.2.1 AVR Simulator *187*
    8.2.2 AVR Studio *188*
8.3 Evaluation Boards *188*
    8.3.1 Atmel AVR MCU00100 Development Board *189*
    8.3.2 STK200 Board *189*
    8.3.3 STK 300 Board *182*
8.4 ICE200 AVR Emulator *192*
8.5 The Device Programmer *193*
8.6 AVR System Design with Components Off the Shelf (COTS) *194*
    8.6.1 The SimmStick Magic *194*
8.7 Code Development with a High Level Language *195*
    8.7.1 C-AVR: A C Compiler for AVR *195*
    8.7.2 DDS MICRO-C Developers Kit for the AVR *197*
    8.7.3 BasicX: A BASIC Interpreter for the AVR *198*
    8.7.4 BASCOM-AVR: A Basic Compiler for the AVR *198*
    8.7.5 JAVRBasic: Jack's AVR Basic Compiler *198*

Chapter 9 Prototyping Techniques 199

9.1 Why Prototype? *199*
9.2 OK, So You Want to Prototype *200*
9.3 Tools of the Trade *202*
9.4 Steps for Prototyping *203*

Chapter 10 AVR Project 1
Smart Dice: A Dice with an Attitude 207

10.1 At A Glance *207*
10.2 Introduction *207*
10.3 Design Issues: Specifying the Requirement *208*
10.4 Design Description *211*
10.5 Possible Alternatives *212*
10.6 Code Development *213*
10.7 Fabrication *217*
10.8 Testing *218*
10.9 Usage *219*
10.10 Power Consumption *219*
10.11 Adapting the Circuit to an AT90S2343 *220*

Chapter 11 AVR Project 2
A Morse Keyer 223

11.1 At a Glance *223*
11.2 Introduction *223*
11.3 Design Specification *225*
11.4 Design Description *225*
11.5 Possible Alternatives *228*
11.6 Fabrication *228*
11.7 Design Code *228*
11.8 Testing the System *229*

Chapter 12 AVR Project 3
A Simple Dual-Channel Voltmeter

233

12.1 At a Glance   *233*
12.2 Introduction   *233*
12.3 Design Description   *234*
12.4 Usage   *234*
12.5 Fabrication   *235*
12.6 Design Code   *235*

Chapter 13 AVR Project 4
The Ubiquitous Kitchen Timer

239

13.1 At a Glance   *239*
13.2 Introduction   *240*
13.3 Design Description   *240*
13.4 Possible Alternatives   *241*
13.5 Fabrication   *241*
13.6 Design Code   *242*
13.7 Testing   *242*

Chapter 14 AVR Project 5
Radio Beacon Controller

245

14.1 At a Glance   *245*
14.2 Introduction   *245*
14.3 Design Specifications   *246*
14.4 Design Description   *246*
14.5 Fabrication   *250*
14.6 Design Code   *250*
14.7 Testing   *252*

Chapter 15 AVR Project 6
AstroDat: A Stand-Alone Data Acquisition System

255

15.1 At A Glance   *255*
15.2 Introduction   *255*
15.3 Design Description for the SniffStick   *257*
15.4 Using the SniffStick   *260*
15.5 AstroDAT: A Complete DAS for Astronomical Application   *261*
15.6 AstroDAT User Interface   *261*
15.7 Design Description   *263*
15.8 System Development   *267*
15.9 Fabrication   *268*
15.10 Design Code   *268*
15.11 Data Readout   *268*
15.12 AstroDat User's Guide   *270*

Chapter 16 AVR Project 7
Security Dongle

277

16.1 At a Glance   *277*
16.2 Introduction   *278*
  16.2.1 What Are Security Locks?   *278*
  16.2.2 Various Hardware Lock Schemes   *278*

16.3 How to Build an Electronic Lock   *280*

16.4 Design Description   *284*

16.5 Possible Alternatives   *286*

16.6 Fabrication   *288*

16.7 Design Code   *288*

16.8 Testing   *289*

Chapter 17 AVR Project 8

A Pulse Frequency Counter with an RS-232 Interface   *291*

17.1 At a Glance   *291*

17.2 Introduction   *291*

17.3 How Does a Frequency Counter Work?   *292*

17.4 How Does a Period Counter Work?   *293*

17.5 Design Description of an AVR-Processor-Based Frequency Counter   *295*

17.6 Usage   *298*

17.7 Fabrication   *298*

17.8 Design Code   *298*

17.9 Testing   *299*

Chapter 18 AVR Project 9

Sa-Re-Ga Follow Me: A Musical Toy   *301*

18.1 At a Glance   *301*

18.2 Introduction   *301*

18.3 Design Description   *303*

18.4 Fabrication   *303*

18.5 Design Code   *305*

Chapter 19 AVR Project 10

AVR Protoboard™ for Nuts™   *309*

19.1 At a Glance   *309*

19.2 Introduction   *309*

19.3 Design Description   *310*

Chapter 20 Ideas for Projects   *317*

20.1 AT90S2343 Controller Based Code Authenticator   *317*

20.2 A CCD Camera Controller   *318*

20.3 Personal Temperature Logger   *318*

    20.3.1 Configuring the Temperature Logger   *320*

    20.3.2 Extracting Data   *320*

20.4 Swipe Card Reader   *320*

20.5 IBM PC Keyboard Decoder   *321*

20.6 A Morse Code Tutor   *321*

Glossary   *325*

Internet Resources for the AVR   *331*

Index   *333*

LIST OF FIGURES

1.1 A digital circuit implemented using TTL ICs.   4
1.2 The digital circuit in Figure 1.1 implemented using a PLD.   4
1.3 An AVR microcontroller-based implementation for the logic equation.   5
1.4 Timing diagrams.   7
1.5 More timing diagrams.   8
1.6 And some more timing diagrams.   9
2.1 A microcontroller interfaces to external devices with a minimum of extra components.   12
2.2 An 8-bit microcontroller.   14
2.3 The ultimate microprocessor development system. The processor accepts binary files through the brain waves in a configurable format!   19
2.4 A more realistic and practical microcontroller development system.   19
3.1 Some AVR controllers.   22
3.2 AVR processor architecture.   23
3.3 AVR processor memory map.   25
3.4 AVR register file.   26
3.5 Instruction fetch/decode and instruction execution.   27
3.6 ALU execution consisting of register fetch, execute, and write back.   28
3.7 On-chip SRAM data access cycles.   28
3.8 The Processor STATUS register.   29
3.9 The general interrupt mask register.   30
3.10 The general interrupt flag register.   30
3.11 The MCU general control register.   31
3.12 The MCU status register.   31
3.13 The Timer/CounterO control register.   33
3.14 The Timer/CounterO register.   33
3.15 The Timer/Counter1 control RegisterA.   34
3.16 The Timer/Counter1 control RegisterB.   35
3.17 The Timer/Counter1 register.   35
3.18 The Timer/Counter1 output compare RegisterA.   36
3.19 The Timer/Counter1 output compare RegisterB.   36
3.20 The Timer/Counter1 input capture register.   37
3.21 The watchdog timer control register.   38
3.22 The EEPROM control register.   39
3.23 The SPI data register.   40
3.24 The SPI status register.   40
3.25 The SPI control register.   41
3.26 The UART I/O data register.   41
3.27 The UART status register.   42
3.28 The UART control register.   43
3.29 The UART baud rate register.   43
3.30 The analog comparator control and status register.   44
3.31 Details of one of the port bits (PORTD4).   46
3.32 Connecting external SRAM to the AVR controllers.   47
3.33 External SRAM to the AVR controller access cycle without wait states.   48
3.34 External SRAM to the AVR controller access cycle with additional wait states.   48

3.35 A clock prescaler for Timer0 as well as Timer1.   *49*

3.36 Timer/Counter0 block diagram.   *49*

3.37 Timer/Counter1 block diagram.   *50*

3.38 UART transmitter block diagram.   *51*

3.39 UART receiver block diagram.   *52*

3.40 Nested interrupt execution.   *54*

3.41 Watchdog timer block diagram.   *56*

3.42 Current consumption by a Tiny22 processor in internal oscillator mode during the active and power down mode.   *57*

4.1 Direct single register access.   *60*

4.2 Direct double register access.   *61*

4.3 Direct I/O memory access.   *62*

4.4 Direct data memory access.   *63*

4.5 Indirect data memory access.   *63*

4.6 Indirect program memory instructions.   *64*

4.7 Relative program memory instructions.   *64*

5.1 A minimum configuration AVR circuit.   *82*

5.2 A rectifier and filter unit.   *83*

5.3 A super simple power supply circuit for the AVR processor powered by the RTS signal pin of the RS-232 port and using a zener diode.   *86*

5.4 Crystal oscillator.   *88*

5.5 Ceramic resonator.   *88*

5.6 Ceramic resonator connected to the oscillator pins of the AVR processor.   *89*

5.7 A quartz crystal connections to the oscillator pins of the AVR processor.   *89*

5.8 Oscillator startup using a parallel resonant crystal after the supply input is applied to the processor.   *90*

5.9 Circuit schematic for the 32KHz clock crystal test circuit.   *91*

5.10 32-kHz oscillator start-up time.   *91*

5.11 Current consumption by an AT90S1200 processor when operated with a 32-kHz clock crystal.   *92*

5.12 Circuit to measure the oscillator frequency variation as a function of supply voltage.   *93*

5.13 Variation of RC system clock frequency as a function of supply voltage.   *93*

5.14 A simple reset circuit.   *95*

5.15 RST signal and the start of the program execution on an AVR processor.   *95*

5.16 Using the DS1233 with an AVR processor.   *96*

6.1 A simple introductory circuit to light a LED.   *98*

6.2 Controlling LEDs with switches.   *99*

6.3 Signal bounce on a mechanical switch when it is released.   *101*

6.4 Connecting AT90S8515 to a PC serial port. Other components that go with MAX232 are not illustrated.   *107*

6.5 Timing the RS-232 signal. The first bit is the Start bit and the last bit is the Stop bit.   *109*

6.6 8-bit digital input port using a parallel-in serial-out shift register.   *110*

6.7 8-bit digital output port using a serial-in parallel-out shift register.   *111*

6.8 8-bit bidirectional digital I/O port expander.   *112*

6.9 AVR interface to PCF8574.   *113*

6.10 An analog signal being sampled and encoded by an ADC. The number output of the ADC is on the Y axis and the time is on the X axis.   *114*

6.11 Analog comparator block diagram.   *114*

6.12 Block diagram for a crude analog-to-digital converter using the on-chip comparator on an AVR processor.   *115*

6.13 A linear and an exponential plot for a small input range. This plot gives an idea of the amount of nonlinearity between the count accumulated using the simple RC charging scheme and the ideal count.   *115*

6.14 Block diagram for an improved analog-to-digital converter using the on-chip comparator on an AVR processor.   *116*

6.15 Block diagram for a temperature sensor interface to the comparator-based ADC.   *117*

6.16 Block diagram of MAX186 ADC.   *118*

6.17 MAX186 control byte format.   *120*

6.18 Timing diagram of a typical MAX186 conversion process as recorded on a logic analyzer.   *120*

6.19 Circuit schematic for an AT90S2313 processor interface to the MAX186 ADC.   *122*

6.20 MAX111 interface to the AVR processor.   *123*

6.21 Timing diagram of the conversion and readout process of the MAX111.   *125*

6.22 A continuously varying PWM signal. The average value of the signal changes by 25% in each period.   *126*

6.23 PWM DAC using an AT90S2313 and an output RC filter.   *126*

6.24 R-2R ladder DAC implementation with an AVR controller.   *127*

6.25 Block diagram of MAX521 DAC.   *128*

6.26 Communication format for MAX521 serial DAC. All transmission begins with a START condition and ends with a STOP condition.   *130*

6.27 Structure of the address and command bytes.   *131*

6.28 Connecting multiple MAX521s on a single bus.   *133*

6.29 Connecting AT90S2313 AVR processor to MAX521 DAC.   *134*

6.30 Seven-segment LED Display Interface to the AVR processor.   *134*

6.31 A Multiplexed Seven Segment LED Display Interface to the AVR processor.   *135*

6.32 An alphanumeric LED display.   *135*

6.33 Block diagram for a 5-x-7 dot-matrix display to AVR interface.   *136*

6.34 Circuit schematic for a 5-x-7 dot-matrix display interface.   *137*

6.35 A 5-x-7 dot matrix display test board photograph.   *138*

6.36 LCD character codes.   *139*

6.37 Circuit schematic for an AT90S2313 processor interface to a 2-line, 16-character LCD.   *140*

6.38 ULN2003A darlington array.   *141*

6.39 ULN2003A drivers used to drive inductive loads.   *142*

6.40 A stepper motor sequencer and driver interface to AVR.   *142*

6.41 Circuit schematic for a stepper motor sequencer and driver for the AVR processors.   *143*

6.42 Ramping the stepper motor speed.   *144*

6.43 EEPROM device address.   *144*

6.44 EEPROM write byte.   *145*

6.45 EEPROM current address read.   *146*

6.46 EEPROM random read.   *147*

6.47 Circuit schematic for an AT90S2313 processor interface to a serial EEPROM.   *148*

6.48 RTC interface to an AT90S2313.   *148*

6.49 Circuit schematic for an AT90S2313 processor interface to an RTC.   *149*

6.50 An arbitrary waveform example.   *151*

6.51 An arbitrary waveform generated by the AVR processor and captured on a logic analyzer. *151*

6.52 A bubble diagram description of a state machine. *154*

7.1 Communication link for AVR processor. The figure illustrates the processor in a point-to-point communication link to another device as well as a link with a bus configuration with multiple devices connected onto the bus. *158*

7.2 How the data is reorganized and extra bit attachments added to the original bit sequence in asynchronous serial data transmission. *159*

7.3 Voltage levels on the RS-232 serial transmission. The waveform is illustrated without any parity bit and 1 stop bit. *159*

7.4 Connecting an AVR device to another AVR or any other serial device. *160*

7.5 Connecting an AVR device to another AVR device using an RS-422 link. *162*

7.6 Original data and the corresponding differential outputs of an RS422 driver. *162*

7.7 Connecting multiple AVR devices on a RS-485 bus. *163*

7.8 Data write and read on an SPI bus. *164*

7.9 IIC bus application. *165*

7.10 Bit transfer on an IIC bus. *165*

7.11 START and STOP conditions on an IIC bus. *166*

7.12 The details of the PC parallel port. *167*

7.13 The DATA port. *168*

7.14 The STATUS port. *169*

7.15 The CONTROL port. *170*

7.16 Connecting an AVR processor to the PC parallel port. *172*

7.17 ISA bus interface for the AVR. *173*

7.18 ISA bus interface data transfer protocol for the AVR. *174*

7.19 ISA bus signals. *175*

7.20 ISA bus signals during Port Read operation. *176*

7.21 ISA bus signals during Port Write operation. *176*

7.22 USB connectivity. *177*

7.23 USB cable. *178*

7.24 USB topology. *179*

7.25 A hub. *179*

7.26 USBN9602 block diagram. *180*

7.27 USBN9602 interface to AT90S8515 AVR controller. *180*

7.28 IrDA physical layer block diagram. *181*

7.29 RZI data encoding scheme employed by IrDA data link. *182*

7.30 An AVR processor with an IrDA data link. *182*

7.31 An AVR processor interface to MAX3100. *183*

7.32 CAN bus topology and signals. *183*

8.1 Windows version of the AVR assembler. *186*

8.2 The AVR simulator. *188*

8.3 The AVR studio. *189*

8.4 Photograph of the MCU00100 evaluation board. *190*

8.5 AVRPROG primary window. *190*

8.6 AVRPROG advanced window. *191*

8.7 AVR ISP software. *191*

8.8 Photograph of the STK200 evaluation board. *193*

8.9 DT104 schematic. *196*

8.10 DT104 component overlay. *197*

8.11 Fully populated DT104 board. *197*

9.1 Photograph of a protoboard.   *201*
9.2 Phtograph of a general purpose printed circuit board.   *201*
9.3 Some useful tools.   *202*
9.4 Component site photograph of a prototype under fabrication.   *204*
9.5 Solder side photograph of a prototype under fabrication.   *205*
10.1 Output LED arrangement for our dice.   *208*
10.2 LEDs light up in this fashion for the numbers 1 to 6.   *209*
10.3 Block diagram for the electronic dice circuit.   *210*
10.4 Schematic for the electronic dice.   *211*
10.5 A typical switch connection configuration for connecting to processors or a digital circuit.   *212*
10.6 Block diagram for an alternative electronic dice circuit. R is a current-limiting resistor for each of the segment LEDs of the 7-segment display.   *213*
10.7 Block diagram for another alternative electronic dice circuit. R is a current limiting resistor for each of the LEDs. The 7 LEDs are again arranged as in Figure 10.1.   *214*
10.8 Photograph of the completed dice circuit board.   *217*
10.9 Photograph of the solder side of the dice circuit board.   *218*
10.10 Possible sources of supply voltage for the dice circuit.   *220*
10.11 Block diagram for the electronic dice using an AT90S2343.   *221*
11.1 Block diagram of the Morse keyer.   *226*
11.2 Circuit schematic for the Morse keyer.   *227*
11.3 Oscillogram for Morse code output for DOT, generated by the keyer circuit.   *230*
11.4 Oscillogram for Morse code output for DASH, generated by the keyer circuit.   *230*
11.5 Oscillogram for Morse code output for the character U, generated by the keyer circuit.   *231*
12.1 Block diagram for the dual-channel voltmeter with LCD.   *234*
12.2 Circuit schematic for a dual-channel voltmeter with an LCD display.   *236*
12.3 Photograph of the dual-channel voltmeter.   *237*
12.4 Logic analyzer screen capture of the MAX111 ADC readout by the AT90S2313 controller.   *237*
13.1 Block diagram of the simple kitchen timer.   *240*
13.2 Circuit schematic for the kitchen timer.   *241*
13.3 Photograph of the kitchen timer.   *242*
13.4 Photograph of a pair of thumbwheel switches used with the kitchen timer.   *243*
14.1 Block diagram of a radio beacon.   *246*
14.2 Morse output for my callsign VU2NOX.   *248*
14.3 Block diagram of a radio beacon controller using the Tiny22 processor.   *249*
14.4 Flowchart for the beacon controller program.   *250*
14.5 Circuit schematic for the radio beacon controller.   *251*
14.6 Photograph of the beacon circuit board.   *252*
14.7 Scope trace for the audio sidetone as well as the transmitter key switch output generated by the beacon controller. The trace shows 4 morse codes for the characters C Q C Q.   *253*
15.1 Using a PC and an external data acquisition system for recording data.   *256*
15.2 An autonomous data acquisition system.   *257*
15.3 Block diagram of SniffStick.   *258*
15.4 Block diagram of the PC parallel-port-based docking port for the SniffStick DAS.   *258*
15.5 Circuit schematic for the SniffStick.   *259*

15.6 Photograph of the SniffStick under fabrication. *260*

15.7 Block diagram of the AstroDat data acquisition system. *262*

15.8 Circuit schematic for AstroDat data acquisition system. *264*

15.9 Signals illustrate the AVR processor controlling the MAX186 ADC. *265*

15.10 Signals illustrate the AVR processor controlling the Dallas DS1302 RTC. *265*

15.11 Signals illustrate the AVR processor controlling the Atmel At24C512 EEPROM. *266*

15.12 Sample data plot of a sinewave generated by a function generator and recorded by the AstroDat System. The X axis is time in ms and the Y axis is volts. *267*

15.13 Completed AstroDat circuit board inside a plastic enclosure. *268*

15.14 Format for the various tags. *269*

15.15 Flowchat for user interaction using the two keys: Acquire and Menu. *271*

16.1 A security lock on the RS-232 port of the PC. *279*

16.2 A security lock on the PC parallel port with a pass-through port. *280*

16.3 An 8-bit linear feedback shift register with taps at bit positions 1, 2, 3 and 7. *281*

16.4 Block diagram of the lock and the PC parallel port signal configuration. *282*

16.5 A 1-input AND gate used as a level isolation circuit. *283*

16.6 The effect of driving a 1-input diode AND gate with a logic signal. *283*

16.7 Circuit schematic for the PC parallel-port-based security lock using AT90S2343. *284*

16.8 Data transfer between a Master and a Slave using Strobe and Ack handshake lines. *285*

16.9 Scope trace illustrates the time relationship between the strobe generated by the PC as the Master and the Ack by the AT90S2343 as a slave. *286*

16.10 Scope trace shows 8 bits of data setup by the PC while sending to the processor and returning data generated by the processor. *287*

16.11 Scope trace shows how the processor can delay the data transfer back to the PC by asserting the Ack signal. When Ack signal remains<+1+>, the PC waits for it to go<+0+> before asserting the Strobe signal to<+1+>. *287*

16.12 Circuit schematic for the PC RS-232 serial-port-based security lock using AT90S2323. *288*

16.13 Photograph of the security dongle. *289*

16.14 A case of a bad power supply with the potential to destroy a 5-V rated processor like the AT90S2343. The trace illustrates the output voltage surging to +12 ?? when it is switched off. *290*

17.1 A frequency counter. *292*

17.2 Frequency counter timing diagram. *293*

17.3 A period counter. *294*

17.4 A period counter timing diagram. *294*

17.5 A compact multifunction period/frequency counter. *295*

17.6 An AVR-based frequency counter with an RS-232 interface. *296*

17.7 Circuit schematic for the frequency counter with an RS-232 interface. *297*

17.8 Logic analyzer trace of the data transmitted by the frequency counter to the PC and the input frequency to the frequency counter. *297*

17.9 Logic analyzer trace of the data transmitted by the PC RS-232 port on the TxD pin and rectified and clamped to convert to unipolar, TTL-level signal on the PB1 pin of the frequency counter. *298*

17.10 User interface for the frequency counter. *299*

17.11 Plot of the input frequency and the measured frequency of the frequency counter. *300*

18.1 Block diagram of this musical toy.   *302*
18.2 Circuit schematic for the toy.   *304*
18.3 A Digital oscilloscope trace of the tone generated by the toy.   *307*
19.1 AVR Protoboard(+tm+) for Nuts(<+tm+).   *310*
19.2 Connecting the AVR protoboard to the PC for program download using the ISP port.   *311*
19.3 AVR protoboard circuit schematic.   *312*
19.4 Photograph of the completed AVR protoboard.   *313*
19.5 Printer port dongle to program the AVR protoboard.   *314*
19.6 Photograph of the printer port dongle to connect the AVR protoboard to the PC.   *315*
20.1 Code authenticator.   *318*
20.2 Block diagram of a CCD camera controller.   *319*
20.3 CCD camera connectivity to the PC.   *319*
20.4 A personal temperature logger.   *320*
20.5 A swipe card reader.   *321*
20.6 A PC keyboard interface to the AT90S2313.   *322*
20.7 A Morse code tutor.   322
20.8 Photograph of the Morse tutor circuit board.   *323*

LIST OF TABLES

3.1 Program memory vector space for AT90S8515. *24*

3.2 Interrupt1 sense control. *31*

3.3 Interrupt0 sense control. *31*

3.4 PORF and EXTRF values after Reset. x means undefined and Y means unchanged. *32*

3.5 ClockO prescale selection. *32*

3.6 Comparel mode select. x is A or B. *33*

3.7 PWM mode select. *33*

3.8 Watchdog timer prescale select. *38*

3.9 SCK frequency. Fcl is the processor oscillator frequency. *41*

3.10 ACIS1, ACISO settings. *44*

3.11 AVR controller selection table. *57*

5.1 MCR control values for DTR and RTS signal voltages. *84*

5.2 RTS voltage variation as a function of load. *85*

5.3 A selection of micropower voltage regulators. *87*

5.4 32-kHz oscillator startup times and current consumption for various capacitor and resistor values. *92*

5.5 Variation of internal RC oscillator frequency with supply voltage. *94*

6.1 ADC MAX186 signals and their functions. *119*

6.2 Signal description of the MAX521 DAC. *129*

6.3 Bits of the command byte for MAX521. *132*

6.4 State transition table. *153*

6.5 State output table. *153*

7.1 Some RS-232 line driver and receiver ICS. *161*

7.2 RS-232 signals and connector pinouts. *161*

7.3 The signals of the Centronics parallel printer adapter. *171*

ACKNOWLEDGMENTS

I acknowledge the following people for their help in completing this book.

David Lee, Ingar Fredriksen, Jarle Boe (all from Atmel), Jack Tidwell, Jason Taylor (Kanda System), Pramod Ranade (SPJ Systems), Pravin Chordia, Rajaram Kharoshe, Saurabh Jain, Smita Mohan, Don McKenzie (Dontronics), Shyam N. Tandon, Premkumar, Vilas Mestry, Sunu Engineer.

This book has been possible because of the encouragement and advice of Scott Grillo (Editor in Chief) and the excellent support from his staff at the McGraw-Hill Technical Book Group.

I have pleasure in acknowledging the patience, support, and encouragement of my wife, Sangeeta, and our son, Chaitanya, during all this time.

My family members Aai, Nana, Sumedha, Sadukaka and Sudhakaku, Appakaka and kaku, and my parents-in-law constantly encouraged me and I thank them all.

Dhananjay V. Gadre
Pune, India

PROGRAMMING AND CUSTOMIZING THE AVR MICROCONTROLLER

INTRODUCTION

This book is about the Atmel's AVR RISC microcontroller series. It covers architecture, design, and usage of this controller in various sample applications. Atmel Corporation (www.atmel.com) is a leading manufacturer of integrated circuits (ICs). AVR is the name of a microcontroller series that Atmel produces and that is the subject of this book. RISC (Reduced Instruction Set Computer) is a popular architecture for modern processors (more about RISC in a later chapter).

Before we get into the details, let us see why it is important to learn about microcontrollers in general and the AVR RISC series in particular. A recent white paper by Sun Microsystems, on picoJava Microprocessor core architecture claims that an average home, by the end of the decade, will contain between 50 to 100 microcontrollers controlling digital phones, microwave ovens, VCRs, televisions sets and television remotes, dishwashers, home security systems, PDAs, etc. Even though this may only reflect the position of a typical home in the advanced countries, there is no denying that even this reflects a huge volume of the microcontroller and microprocessor usage in the home environment. Besides home use, another area that is fueling the microcontroller growth is electronic commerce. With the advent of "smart cards," which have much more storage capacity than the more conventional magnetic cards and are more reliable, these devices are all set to replace paper currency, which means that a humongous number of people will be using the smart cards. There is even more: An average car has about 15 processors; the 1999 Mercedes S-class car has 63 microprocessors, while the 1999 BMW has 65 processors! In fact, except perhaps the human body, microprocessors and microcontrollers have gotten into everything around us (and even that may not be completely true—it would not be surprising if a heart pacemaker is microprocessor controlled).

Microcontrollers or microprocessors are easier to use as a controller than say a dedicated digital state machine in a system such as a washing machine, for example, cheaper to upgrade, and require less inventory; all issues critical for maintaining economic viability and profit in the face of cutthroat competition—thus this great rush for microcontrollers and microprocessors. Given the mass usage of microcontrollers in devices, systems, and consumer components, it is obvious where the money is.

So we want to learn about microcontrollers and microprocessors. However, you might have noticed that I have used these terms interchangeably and rather loosely. It is time to consider what a microcontroller really is and how it differs from a microprocessor.

1.1   Microcontroller, Microcomputer, or Microprocessor?

It is common to hear these terms being used interchangeably. However, each is quite distinct from the other and it is important to understand the differences at this point.

A microprocessor is a central processing unit (CPU) on a single chip. In the olden times, the CPU was designed using many medium/large scale integrated (MSI, LSI) chips. Intel, with its 4004, put all the components of a CPU—arithmetic logic unit (ALU), instruction decoder, registers, bus control circuit, etc.—on a single chip, and so the microprocessor was born. The 4004 was a 4-bit (i.e., it processed data in chunks of 4 bits at a time) microprocessor designed to be the number cruncher in a calculator.

When a microprocessor and associated support circuitry, peripheral I/O components and memory (program as well as data) were put together to form a small computer specifically for data acquisition and control applications, it was called a microcomputer.

So if I were to design a circuit with a popular microprocessor 8088 or for that matter even the 8085, put in EPROM for storing the program, RAM for storing variables and results and a few I/O interface chips for interacting with the external world, I would have put together a microcomputer.

In a logical extension, when the components that make a microcomputer were put together on a single chip of silicon, it was called the microcontroller. Texas Instruments is credited with creating the first microcontroller, the TMS1000 series. The TMS1000 series microcontrollers had enough RAM, ROM, and I/O and were used as microwave oven controllers, in industrial timers, and in calculators.

Today there are many microcontroller families: Intel's 8048 and 8051, Motorola's 68HC11, Zilog's Z8, Microchip's PIC, Hitachi's H8, and now Atmel's AVR. A microcontroller family indicates the availability of many different microcontrollers with the same basic central core but different peripherals, packaging, operating speed options, etc.

Even though the definitions for a microprocessor, a microcomputer, and a microcontroller are clear and unambiguous, it is quite common to see these terms being used loosely and interchangeably. This fuzziness in terms exists and we will have to live with it. For our work we will use the term microcontroller for a chip with on-chip memory and peripheral I/O capability (ports, timers, serial port, etc.) besides the CPU.

The Atmel's AVR controller, with its on-chip program and memory, I/O ports, timers, and serial port, is a microcontroller, as it certainly satisfies the above criteria.

1.2   Do You Need a Microcontroller?

Looking at the needs, decide whether it can be done simply. It requires substantial investment of time, money, and effort to put together a reliable microcontroller-based system. The advantages are small overheads when upgrading the system with small changes. It also helps to keep the inventory to a relatively small number of components.

Possible alternatives are:

1. A dedicated digital circuit,
2. A digital circuit based on a PLD (programmable logic device),
3. An application specific integrated circuit (ASIC) based implementation.

The above-mentioned alternatives to microprocessors are quite similar and differ in only the implementation. A dedicated digital circuit might use discrete ICs for the various logic functions (AND, OR, XOR, etc.) while a digital circuit based on a PLD would be more compact given the programmable nature of a PLD. A PLD contains an array of various logic function blocks, the user selects the required functionality, and the interconnection between these functional blocks at the design level, thereby achieving a more integrated and compact solution. A PLD has a substantial amount of hardware, of which only a fraction gets utilized in average applications. The ASIC solution is like a PLD except that it is an optimized implementation.

Figure 1.1 is the circuit diagram for implementing an hypothetical logic equation using individual digital ICs. These logic gates, as seen in Figure 1.1, are available in various logic families (TTL, CMOS, etc.). The figure illustrates the IC numbers for the TTL family. To implement this equation, we need 3 ICs with about 57 percent utilization (the 7404 IC has 6 gates and we have used 3 of them, while the 7408 and 7432 has 4 gates each, of which we have used 5 gates—8 gates in all out of 14 available gates, i.e., a utilization factor of .57). The same equation is now implemented using a PLD (such as 16L8).

Figure 1.2 illustrates the internals of a PLD implementation. Each intersection in the AND array represents an AND gate, while each intersection in the OR array represents an OR gate. For this solution, we only need 1 IC. The PLD in Figure 1.2 has about 150 gates, of which we have used only about 12, representing a mere 8 percent utilization! (The actual 16L8 if used for this circuit has more hardware than seen in Figure 1.2.) A PLD-based circuit is also more power-consuming than a comparable ASIC circuit, which is due to the redundant hardware on the PLD chip.

In contrast, a microprocessor-based (in fact an Atmel AVR processor-based) circuit is illustrated in Figure 1.3. It is as small as the PLD-based circuit, and in terms of power consumption, is better than a PLD circuit. In terms of speed, the PLD will perform much faster than a processor. Of course, for the microprocessor circuit to work correctly, it must

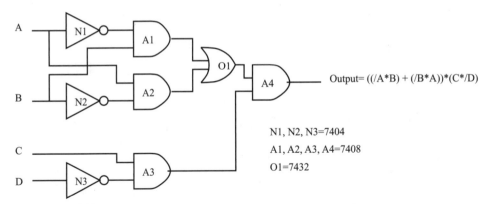

FIGURE 1.1 A digital circuit implemented using TTL ICs.

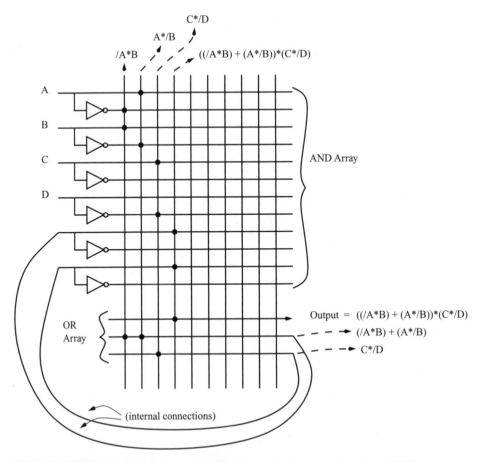

FIGURE 1.2 The digital circuit in Figure 1.1 implemented using a PLD.

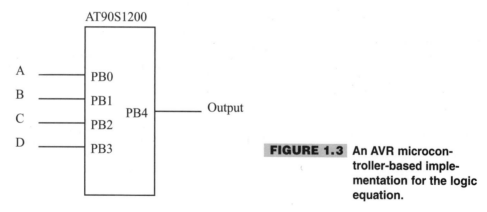

AT90S1200

A —— PB0

B —— PB1

         PB4 —— Output

C —— PB2

D —— PB3

FIGURE 1.3 **An AVR microcontroller-based implementation for the logic equation.**

be programmed correctly. The program to implement our logic equation is discussed in a later chapter.

While we are trying to portray the microprocessor circuit in a positive light, it is worthwhile to be able to remember the relative merits and demerits of each implementation. It is not that the microprocessor is the solution to all problems. Sometimes you need to use a PLD in conjunction with a processor, and sometimes a PLD alone is required. One such implementation of a processor and a PLD working together is the subject of a later chapter.

1.3   Why the Atmel's AVR Microcontroller?

Whether a particular requirement needs to be implemented using discrete ICs or PLDs or a microprocessor must be determined by the designer. However, many applications could be suitably implemented using microcontrollers, and a great many of them would benefit from using the AVR as outlined briefly below.

We will discuss the AVR features in detail in later chapters, but at this point it may be useful to outline the salient features. Atmel's AVR RISC family of controllers has the following features:

1. RISC architecture with mostly fixed-length instruction, load-store memory access, and 32 general-purpose registers.
2. A two-stage instruction pipeline that speeds up execution.
3. Majority of instructions take one clock cycle.
4. Up to 10-MHz clock operation.
5. Wide variety of on-chip peripherals, including digital I/O, ADC, EEPROM, Timer, UART, RTC timer, pulse width modulator (PWM), etc.
6. Internal program and data memory.
7. In-system programmable.
8. Available in 8-pin to 64-pin package size to suit wide variety of applications.
9. Up to 12 times performance speedup over conventional CISC controllers.

10. Wide operating voltage from 2.7 V to 6.0 V.
11. A simple architecture offers a small learning curve to the uninitiated.

What does the name AVR stand for? Atmel says that it is just a name. However, AVR seems to have the initials of the people who designed the controller.

1.4   Organization of This Book

The book has three logical sections:

1. Introduction and preliminary discussion about microcontrollers and AVR controller details. These are covered in Chapters 1 to 4.
2. System design using the AVR RISC controllers. Issues include system design, code development, software and hardware interfacing the AVR to the outside world. These aspects are covered in Chapters 5 to 9.
3. Sample applications are covered in the rest of the chapters, and these illustrate how the AVR controller could be used in real applications.

The idea is to present the material in a format that is easily accessible to readers of varying interests. Beginners could start from the initial chapters and work their way up till the very end. An individual with some experience with microcontrollers could, on the other hand, skip the initial chapters in Section 1 and pick up where new material is presented. However, the last section on applications could be a starting point for beginners as well as experienced users to give perspective. The sample applications illustrate the various ways in which this versatile family of controllers could be used and could well be a starting point for a beginner.

The middle sections deal with the specifics of the AVR controller family and how to get an application up and running, how to develop code, and the various tools available in the form of assemblers, compilers, simulators, evaluation and prototyping boards. I have sampled a few of these commercial and freeware offerings, and I present my opinion about these products in these sections.

1.5   Timing Diagram Conventions

Timing diagrams are the key to understanding digital circuits and systems. Timing diagrams illustrate how the signals of a circuit vary as a function of time, as well as the interplay between the signals. They are the starting point in describing the way a circuit or a system ought to work, and after a circuit has been designed, the timing diagrams tell the readers how the circuit or the system works. In turn, if this circuit is being used as a subsystem in a larger design, its associated timing diagram will determine how it fits into the larger system or how the larger system is to be designed to accommodate the smaller subcircuit. (Figure 1.4.)

1. A low-level signal is zero voltage (actually a range of voltage around zero), and a high-level signal is supply voltage (or a range around it).

2. Transition of a low-level signal to high level.

3. Transition of a high-level signal to a low level.

4. Transition of a bunch of parallel signals (called BUS) from one level to another.

5. A signal that goes in a high-impedance state, also called a floating signal.

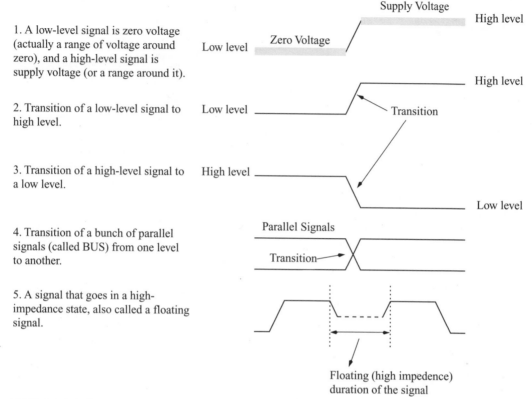

FIGURE 1.4 Timing diagrams.

Together with a circuit diagram, it is the electrical engineer's equivalent of an architectural plan of a building. A circuit diagram and the associated timing diagram completely and exactly describe the circuit's working. (Figures 1.5, 1.6.)

To understand circuits and the timing diagrams, we must follow a uniform convention. To describe the various states a digital circuit operates in, we have a number of symbols. This section defines the conventions we will follow in this text.

6. A BUS with floating signals.

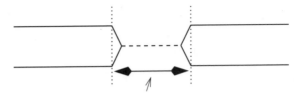

Duration of the signal when it is
in high impedance, floating state

7. A change of condition on one signal
causes a transition on another signal.
The example shows a high-to-low signal
transition causing a high-to-low-level
transition on another signal.

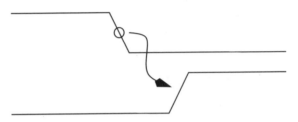

8. A transition on a signal causes a
transition on a BUS.

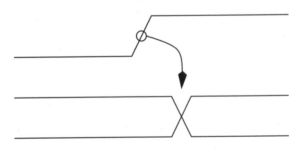

FIGURE 1.5 More timing diagrams.

9. More than 1 condition must exist to force a transition of signals on the BUS. Example shows that a transition on one signal during the time when the other signal is at high level causes a transition on the BUS signals.

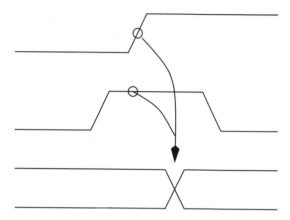

10. A condition on a signal causes changes on more than one signal level. The example shows that a high-to-low transition on one signal causes a high-to-low transition on the second signal and a pulse on another signal.

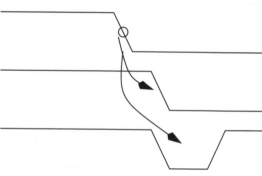

FIGURE 1.6 And some more timing diagrams.

$$2$$

MICROCONTROLLERS

In this chapter we will briefly outline the history of microcontrollers and then partition the various devices in different categories. But before that, let's take a fresh look at what a microcontroller is and what can it do for us.

Microcontrollers are fun. They are the heart and soul of many everyday appliances. And most of all, microcontrollers are easy to use and to design with, from the point of view of a designer. Figure 2.1 is the block diagram of what a typical modern microcontroller, and especially those in the AVR series, can do. The block in the center of the figure represents the microcontroller. It can interface to motors, a variety of displays as output devices, communicate to PCs, read external sensor values, even connect to a network of similar controllers, and it can do all that without a lot of extra components. This leads to a small and compact system that is more reliable and cost-effective (because of the fewer number of components and the fewer number of interconnections).

Contrast that with a situation where you don't have the microcontroller: You only have a CPU. To build a system to interface to various devices (motors, displays, etc.) you would need external program memory and RAM besides the other required peripheral interface components needed to connect the motors, displays, sensors, etc., to the CPU chip. Imagine the number of additional components! Rather than being a single-chip system, you would end up with a system with a board full of components with increased power consumption.

Let us look at the various microcontroller components.

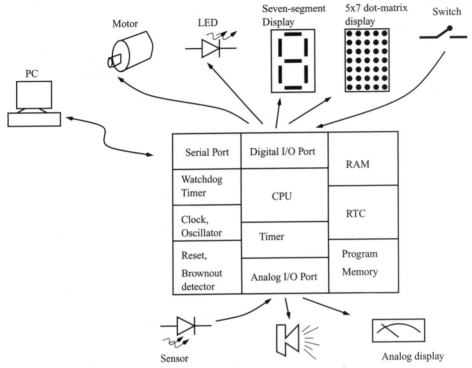

Figure 2.1 A microcontroller interfaces to external devices with a minimum of extra components.

1. **CPU:** The central processing unit (CPU) is the heart of the controller. It fetches the instructions stored in the program memory, decodes these instructions, and executes them. The CPU itself is composed of registers, the arithmetic logic unit (ALU), instruction decoder, and control circuitry.

2. **Program Memory:** The program memory stores the instructions that form the program. To accommodate larger programs, the program memory may be partitioned as internal program memory and external program memory in some controllers. Program memory is usually nonvolatile and is of EEPROM or EPROM or Flash or Mask ROM or OTP (one-time programmable) type.

3. **RAM:** The RAM is the data memory of the controller, i.e., it is used by the controller to store data. The CPU uses RAM to store variables as well as the stack. The stack is used by the CPU to store return addresses from where to resume execution after it has completed a subroutine or an interrupt call.

4. **Clock Oscillator:** The controller executes the program out of the program memory at a certain rate. This rate is determined by the frequency of the clock oscillator. The clock oscillator could be an internal RC-oscillator or an oscillator with an external timing element, such as a quartz crystal, an LC resonant circuit, or even an RC circuit. As soon as the power is applied to the controller, the oscillator starts operating.

5. **Reset and Brownout Detector Circuit:** The reset circuit in the controller ensures that at startup all the components and control circuits in the controller start at a

predefined initial state and all the required registers are initialized properly. The brownout detector is a circuit that monitors the power supply voltage, and if there is a momentary drop in voltage, resets the processor so that the drop in voltage does not corrupt register and memory contents, which could lead to faulty operation of the controller.

6. Serial Port: The serial port is a very useful component on the controller. It is used to communicate with external devices on a serial data basis. The serial port can operate at any required data transfer speed. The serial port takes data bytes from the controller and shifts out the data one bit at a time to the output. Similarly, it accepts external data a bit at a time, makes a byte out of 8 such bits, and presents this to the controller. Serial ports are of two types: synchronous and asynchronous. Synchronous data transfer needs an accompanying clock signal with each data bit for timing information, while the asynchronous data transfer does not need the clock signal, and the timing information and synchronization is embedded in the data bit itself by way of duration of data bits as well as additional start- and stop-bits on the data path.

7. Digital I/O Port: The microcontroller uses the digital I/O components to exchange digital data with the outside world. Compared to the serial port, which transfers data serially one bit at a time, the data on the digital I/O port is exchanged as bytes.

8. Analog I/O Port: Analog input is performed using an analog-to-digital converter (ADC). The controller could be equipped with an integrated ADC or an analog comparator, which is used under software control to perform A-to-D conversion. ADCs are used to acquire sensor data from devices such as temperature sensors and pressure sensors; such sensors often produce proportional analog voltage data. Analog output is performed using a digital-to-analog converter (DAC). Most controllers are equipped with pulse-width modulators that can be used to get analog voltage with a suitable external RC filter. DACs are used to drive motors, for visual displays (of the older VU meter types), to generate sound or music, etc.

9. Timer: The timer is used by the controller to time events; e.g., it may be required to output data to a display at some rate. The timer would be used by the controller to generate that rate. The timer can also be used to count events, external as well as internal. In that case the timer is called a counter.

10. Watchdog Timer: A watchdog timer (WDT) is a special timer with a specific function. It is usually used to prevent software crashes. It works as follows: Once armed, the WDT increments an internal counter at some rate. If the user program does not reset the counter, the counter overflows, which is used to reset the controller. The user software is programmed suitably, therefore, frequently enough, to reset the WDT to give a sort of "I am alive" indication. The assumption is that if the user program does not reset the WDT, it has failed in some way and therefore rather than a system crash or unpredictable system performance, it is better to reset the system.

11. RTC: A real timer clock (RTC) is a special timer with the task of maintaining time of day, date, etc. It can be used to time-stamp events.

While Figure 2.1 illustrates a typical microcontroller, these devices come in a variety of sizes and complexity. Like microprocessors (i.e., CPU on a chip), microcontrollers are classified as 8-bit, 16-bit, or 32-bit (or 64-bit) components. This refers to the width of the internal registers and the accumulator. An 8-bit system usually also means that the CPU

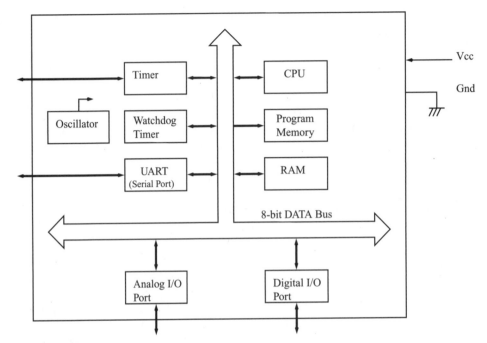

Figure 2-2 An 8-bit microcontroller.

connects to the various chip components through an 8-bit data path. Figure 2.2 illustrates this concept.

Of the various microcontroller types, 8-bit microcontrollers have the largest market share. In 1999, the market for 8-bit chips was $4.8 billion. In comparison, the combined 16-bit and 32-bit chip market was merely $452 million only. Smaller 4-bit controllers also exist and have a small market share.

Controllers with larger data paths can perform better than similar controllers with smaller data paths. However, controllers with smaller data paths also have cheaper development tools compared to controllers with bigger data paths. Eight-bit controllers are the most popular devices not only because of lower device cost (compared to 16- or 32-bit devices) but also because the development tools for 8-bit devices cost much less, and 8-bit devices are now being offered with increased performance and more integrated peripheral components.

Besides the classification based on the size of the internal data path, microcontrollers are also classified on the basis of the underlying architecture. The next section looks at architectural aspects of the microcontroller.

2.1   Microcontroller Architecture

Microcontroller architecture is classified on the basis of various features. One very common classification is on the basis of a number of instructions: CISC (complex instruction set computer), RISC (reduced instruction set computer), or MISC (minimal instruction set computer). However, these terms have been much muddled by marketing personnel. A CISC processor often has many RISC-like features, and it has become very confusing.

Another classification is on the basis of way the program and data memory is accessed; a unified memory model is called the Princeton or Von Neumann architecture versus the Harvard architecture, which offered separate memory for program storage and data storage.

Another classification is on the basis of the way the internal data is stored and manipulated inside the CPU. A microcontroller's job is to manipulate data. A microcontroller (or a microprocessor) manipulates data with the help of a user program. The way this data is stored and accessed internally in the CPU and the way it is processed forms the basis of different processor architectures and yet another classification scheme. There are four basic models: stack, accumulator, register-memory, and register-register (known as load-store).

To understand the differences between these various architectures (on the basis of internal data manipulation) let us consider code sequences for performing the following computation:

$$C = A - B$$

where A, B, and C are variables.

A stack machine performs this computation as follows:

```
Push A
Push B
Sub
Pop   C
```

In a stack machine, the ALU gets all operands from the stack and stores all operands back on the stack. To load a variable on the stack, an instruction Push Var is used. A stack operates by putting the last value on the top. The ALU accesses the top two values on the stack and performs any given operation (addition, subtraction, division, etc.). The result is stored back on the stack at the topmost location.

An accumulator machine performs this computation as follows: In the accumulator machine, one of the operands is always the accumulator. In fact, all operations are accumulator centric.

```
Load A; Loads accumulator with variable A
Sub B; subtracts variable B from the contents of the accumulator and
  stores the result back in accumulator
Store C; stores the value of the accumulator, which has the result, in
  variable C
```

A register-memory machine performs this computation as follows:

```
Load Rx, A; loads a register Rx with variable A
Sub Rx, B; subtracts the variable B from the contents of register Rx
  and stores the result in Rx
Store C, Rx; stores the contents of Rx which is the result in variable C
```

A register-register machine performs this computation as follows:

```
Load Rx, A
Load Ry, B
```

```
Sub Rz, Rx, Ry
Store C, Rz
```

In the register-register model, memory (which stores the variables) is accessed only using the load- and-store instructions. Hence here, the registers are first loaded with the variable values, the computation is performed with the result back in one of the registers, and the result from this register is stored back in the destination variable.

The register-memory and the register-register architecture processors have a large number of registers that are orthogonal in nature. Any register can be used in any operation. Typically, such architectures have 32 general-purpose registers.

Early processor architectures used either the stack or the accumulator model. However, most modern processors use the register-register architecture. This is because of the realization that accessing internal registers is much faster than accessing external memory. To reduce external memory accesses, a large pool of general-purpose registers is provided for the register-register model. Moreover, registers are easier to access for a compiler than say a stack, even though the stack is inside the processor.

2.2   Choosing a Microcontroller

There are literally hundreds of microprocessors and microcontrollers on the market, and choosing a particular one for your application can be a nightmare. Usually one starts by enumerating one's requirements in terms of features and cost and then comparing these with what is available. The final choice may still be dictated by other factors such as market trends, company profile, popularity, local design expertise, etc.

Listed below are some of the popular 8-bit microcontrollers and their features. These devices are the lowest cost-representative devices from respective manufacturers.

COMPANY	DEVICE	ON-CHIP MEMORY	OTHER FEATURES
AB Semicon Ltd	AB180-20	Nil	Two 16-bit timers, UART, fixed-point 32-bit arithmetic unit, DMA controller
Atmel Corp	ATtiny11	1-kbyte flash	8-bit timer, analog comparator, watchdog, on-chip oscillator, one external interrupt
Dallas Semi	DS80C310	256-byte RAM	Four clocks per machine cycle, UART, three 16-bit timer/counters, dual data pointers, ten internal/six external interrupts, power-on reset
Hitachi	H8/3640	8-kbyte ROM 512 byte RAM	Three 8-bit timers, one 16-bit timer, one 14-bit PWM timer, one watchdog, two SCI ports, eight 8-bit ADC, 32-kHz sub-clock generator

COMPANY	DEVICE	MEMORY	OTHER FEATURES
Infineon	C501	8-kbyte ROM 256-bytes RAM	Serial interface, three 16-bit timers, 32 I/O ports
Microchip	PIC16CR54C	768-byte ROM, 25-byte RAM	12 I/O pins, 8-bit timer, high-current sink/source for direct LED drive, watchdog timer, RC oscillator
Mitsubishi	M37531M4	8-kbyte ROM 256-byte RAM	2.2 to 5.5V operation; 16-bit-wide address bus; three 8-bit timers; 16-bit watchdog timer; 10-bit, eight-channel ADC; UART; one clock-synchronized serial port; one external interrupt, seven high-current output ports for LED operation; key-on wake-up function, 29 program-mable-I/O ports, built-in clock-generating circuit
Motorola	68HC705KJ1	1240-byte OTP 64-byte RAM	15-stage multifunction timer, on-chip oscillator, low-voltage reset, watchdog, keyboard interrupt, high-current I/O port
NEC	789011	2-kbyte RAM 128-byte RAM	Two 8-bit timers, UART, 22 pro-grammable I/O ports, two-channel serial interface
Philips	P87LPC762	2-kbyte OTP 128-byte RAM	Oscillator, watchdog, 32-byte customer-code EPROM, UART, I2C, comparators, timers/coun-ters, brown-out detector, power-on reset, keypad wake-up, LED drivers
Samsung	KS86C0004	4-kbyte ROM 208-byte RAM	RC oscillator, 12-pin key matrix, one 8-bit timer, one 8-bit timer/counter, 14 interrupt sources, 32 I/O ports
Scenix	SX28AC	3-kbyte flash 136-byte RAM	Analog comparator, program-mable I/O, brown-out detector, 8-bit timer, watchdog
STMicro	ST6203CB1	1 kbyte ROM or OTP 64-byte RAM	8-bit timer, watchdog, nine I/O lines with high-current capabili-ty, internal backup oscillator system, brown-out detection
Toshiba	TMP87C405AM	4-kbyte ROM 256-byte RAM	Nine interrupt sources, pro-grammable watchdog timer, 22 programmable I/O ports

COMPANY	DEVICE	MEMORY	OTHER FEATURES
Xemics SA	XE8301	22-kbyte ROM 512-byte RAM	Clock prescalar, watchdog timer, power-on reset, supply-level detection, 20-pin programmable I/O, crystal and RC oscillator, UART, four 8-bit timers with PWM
Zilog	Z8E000	0.5-kbyte OTP 32-byte RAM	One 16-bit timer, watchdog, four hardware interrupts, 13 I/O pins

2.3 Developing Applications with a Microcontroller

Now that we have a little bit of inside information about microcontrollers and what can be done using them, it is time to discover how to go about developing applications using these controllers.

An ideal and a rather futuristic method is depicted in Figure 2.3. But let us for a moment consider what all is required to develop applications using controllers.

Let us list one of the possible roadmaps for designing a microcontroller-based device.

1. First and foremost, define the requirements.
2. Create sufficient documentation to support the requirements in the form of block diagrams, flowcharts, timing diagrams, etc.
3. Search for suitable hardware to provide the necessary functionality. This may help the designer realize whether a microcontroller is needed at all or not.
4. If you do need a microcontroller, then identify a suitable microcontroller that can act as the brains for the device.
5. Once you have identified the controller, double-check that in fact the microcontroller will satisfy the requirements in terms of speed, power consumption, etc. Otherwise you will have to iterate once again to choose another controller.
6. As a next step you will need to acquire all the tools to help develop the hardware and the software. These tools may include an assembler and/or a compiler if you wish to program in a high-level language, a simulator for the controller, if possible a hardware emulator, evaluation board, programmer for the controller, etc.
7. If you are already familiar with this particular controller, you can start designing and assembling a prototype; otherwise, you may need to get familiar with the controller by writing sample programs and testing them on the evaluation board or on the software simulator.
8. Once you become familiar with the features of the controller, you can start partitioning the software in manageable blocks that can be written as subroutines and tested

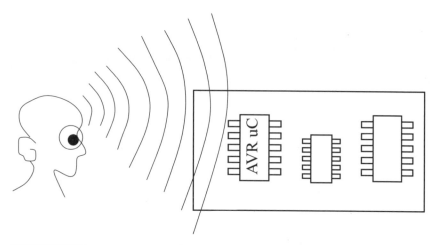

Figure 2.3 The ultimate microprocessor development system. The processor accepts binary files through the brain waves in a configurable format!

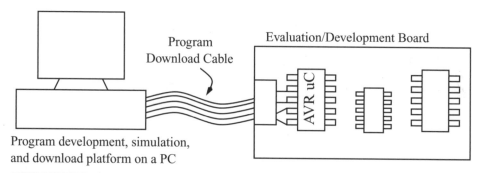

Figure 2.4 A more realistic and practical microcontroller development system.

independently. The hardware development can go on in parallel, and you enter a phase that could be termed as test-and-debug phase. This is an iterative phase that continues till all your subroutines and hardware work as required. To complete this phase, you must take the help of as much test equipment and development tools (such as a programmer and an emulator) as possible to minimize the number of iterations. A realistic development system is illustrated in Figure 2.4.

9. Finally, you must integrate all the software and the hardware and test again. You again enter the test-and-debug phase till everything works as required.

10. During the writing of the software and building the hardware, an important activity that should not be missed is documentation. Documenting the design is extremely important not only for maintaining a record of your work, but also for testing during device lifetime and future revisions.

11. The final stage involves deploying the system in the target environment or the appropriate production line.

3

THE AVR RISC MICROCONTROLLER
ARCHITECTURE

3.1 Introduction

This chapter describes the AVR processor family in detail, covering the architectural aspects and the integrated peripheral components that are bundled with the CPU.

The AVR is a RISC processor with a Harvard architecture. The Harvard architecture refers to the fact that the CPU has a program memory and a separate data memory.

The AVR processor family has the following features:

1. On-chip and In System Programmable Flash memory used as Program Memory. All the processors have on-chip flash program memory. This means you don't have to have external EPROMs or ROMs containing your program code. Also, the program memory can be programmed while the processor is in the target without removing it. This allows faster and easier system software upgrades. The program memory can be programmed in situ (i.e., without removing from the target system). The program memory can be programmed in two modes: serial and parallel, which we will discuss later.

2. 32-X-8 general-purpose working registers (in the true RISC tradition). A large register set means that variables can be stored inside the CPU rather than storing the variables in memory, as accessing memory, is time expensive. Thus the program will run faster.

3. On-chip data memory EEPROM and RAM in most devices. The CPU is Harvard architecture, and the EEPROM and the RAM is seen as DATA memory for storing constants and variables.
4. 0 to 10-MHz clock speed operation. Most instructions operate in 1 clock cycle, and this leads to an almost 10-times performance improvement over conventional processors (e.g., the 8051) operating at equal clock frequency.
5. Power On RESET circuit.
6. On-chip programmable timer with separate prescalar. This is used for timing applications.
7. Internal and external interrupt sources.
8. Programmable watchdog timer with independent oscillator. This is used to recover in case of software crash but can also be used for other interesting applications as discussed in one of the project chapters.
9. SLEEP and POWER DOWN modes of operation. This saves power when the processor is idling.
10. Many chips with on-chip RC clock oscillator. Using the on-chip RC oscillator feature when available leads to an even lower component count.
11. Wide device range (from a small 8-pin processor to a 68 pin processor), and one can choose a processor to suit a given requirement while being able to use the same development facilities. (Figure 3.1)

3.2   AVR Family Architecture

The AVR uses Harvard architecture. This entails separate data and program memory buses. Figure 3.2 illustrates the controller layout. The data memory data bus is an 8-bit bus and connects most of the peripheral components to the register file. The program memory data

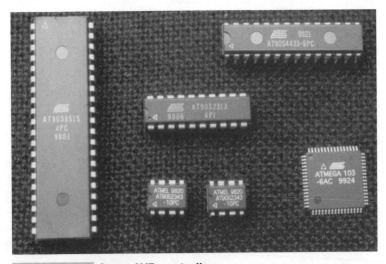

FIGURE 3.1 Some AVR controllers.

bus is 16 bits wide and only feeds the instruction register. Even though Figure 3.2 refers to the AVR AT90S2313 controller, it applies equally well to all the processors and differs only to the extent of having additional (or less) peripheral components as well as differing in amounts of program memory and data memory.

The program memory is a continuous chunk of flash memory. The exact amount varies from processor to processor. The AT90S1200, the base level processor, has 1 Kbyte of program memory organized as 512-X-16 bits, while the Mega103 has 128 Kbytes of memory organized as 64K-X-16 bits. A K here equals 1024 and not 1000. The program memory is accessed every clock cycle, and an instruction is loaded into the instruction register. The instruction register feeds the register file, selecting which of the registers will be used by the ALU for instruction execution. The instruction register output is also decoded by the instruction decoder to decide which control signals will be activated for completing the current instruction.

The program memory, besides storing instructions, also stores interrupt vectors starting at address $0000. The actual program should start at memory location beyond the space meant for the vectors. The number of vectors vary from processor to processor.

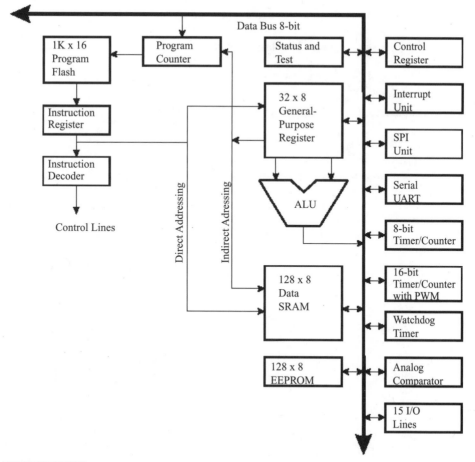

FIGURE 3.2 AVR processor architecture.

AT90S1200 has 3 vectors and AT90S8515 has 13 vectors. Table 3.1 illustrates the complete vector space for AT90S8515.

The data memory, on the other hand, is split up in different types. Figure 3.3 has the various memory maps available to an AVR processor. The data memory has in all five different components:

1. A register file with 32 registers of 8-bit width. All processors of the AVR family have this register file.

2. 64 I/O registers of 8 bits each. All the processors do not have all the 64 registers. Some have more than others, depending on the number of peripheral components on the chip. These registers are really part of on-chip SRAM and can be accessed either as SRAM with addresses between $20 and $5F or as I/O registers with addresses between $00 and $3F. Most often, all of these are accessed as I/O registers rather than as SRAM.

3. Internal SRAM. This is available on most of the AVR processors except the baseline processors such as the AT90S1200. The amount of SRAM varies between 128 bytes to 4 Kbytes. The SRAM is used for stack as well as storing variables. During interrupts and subroutine calls, the current program counter value is stored on the stack. The size of the stack is limited by the available on-chip SRAM. The current stack location is indicated by the stack pointer. The stack pointer is 1 byte on smaller processors such as AT90S2313 and is 2 bytes on larger processors such as the AT90S8515. The stack pointer must be initialized after reset and before the stack can be used. For those processors that do not have on-chip SRAM such as the AT90S1200, a hardware stack is available to store program return addresses. This hardware stack can only store up to 3 return addresses.

TABLE 3-1 PROGRAM MEMORY VECTOR SPACE FOR AT90S8515

PROGRAM MEMORY ADDRESS	VECTOR	COMMENTS
$0000	Reset	Reset handler
$0001	EXT_INT0	IRQ0 handler
$0002	EXT_INT1	IRQ1 handler
$0003	TIM_CAPT	Timer1 capture handler
$0004	TIM1_COMA	Timer1 compareA handler
$0005	TIM1_COMB	Timer1 compareB handler
$0006	TIM1_OVF	Timer1 overflow handler
$0007	TIM0_OVF	Timer0 overflow handler
$0008	SPI_STC	SPI transfer complete handler
$0009	UART_RXC	UART RX complete handler
$000A	UART_DRE	UART UDR empty handler
$000B	UART_TXC	UAT TXC complete handler
$000C	ANA_COMP	Analog comparator handler

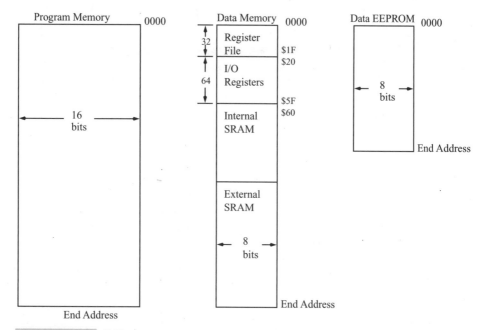

FIGURE 3.3 AVR processor memory map.

4. External SRAM. This is possible only on the larger processors of the AVR family. Those processors that have external data and memory access ports (such as the AT90S8515) can use any available external SRAM the user may decide to implement.

5. EEPROM. The EEPROM is available on almost all AVR processors and is accessed in a separate memory map. The starting address of the EEPROM is always $0000. Various processors have between 64 bytes and 4 Kbytes of EEPROM. The EEPROM can be read and written by any program. Reading the EEPROM is faster than writing the EEROM. The EEPROM can be written to about 100,000 times.

Most of the AVR instructions are 1 word (2 bytes) long and so take 1 program memory location. Many instructions execute in a single clock cycle, and a few take 2 or more clock cycles. This single-cycle execution is achieved due to the use of a 2-stage pipeline. The pipeline works by concurrently acquiring a new instruction from the program memory while the previous instruction is executing in the other part of the processor. Thus instruction fetch and decode and execution are processes that are being performed by the processor concurrently.

Now let us look at the various components that make the AVR processor.

3.3   The Register File

All AVR processors have 32 general-purpose registers. Some of these registers have additional special functions. The registers are named R0 through R31. The register file is broken up into 2 parts with 16 registers each, R0 to R15 and R16 to R31. All instructions that operate on the registers have direct access and single-cycle access to all the registers. The exception is the SBCI, SUBI, CPI, ANDI, and ORI instructions as well as the LDI instruction. These instructions operate only on registers R16 to R31.

Registers R0 and R26 through R31 have additional functions. R0 is used in the instruction LPM (load program memory), while R26 through R31 are used as pointer registers as illustrated in Figure 3.4. These pointer registers are used in many of the register indirect instructions.

3.4   The ALU

The arithmetic logic unit (ALU) performs such operations as bit, arithmetic, and logic upon the contents of the registers and writes back the result into the register file into the designated register. These operations are performed in a single clock cycle. Each ALU operation affects the flags in the STATUS register, depending upon the instruction.

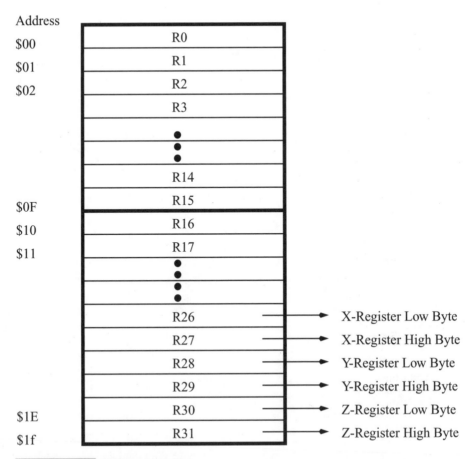

FIGURE 3.4 AVR register file.

3.5   Memory Access and Instruction Execution

The AVR processor is driven by the system clock, which can be sourced from outside or, if available and enabled, an internal RC clock can be used. This system clock without any division is used directly for all accesses inside the processor. The processor has a two-stage pipeline, and instruction fetch/decode is performed concurrently with the instruction execution. This is illustrated in Figure 3.5.

Once the instruction is fetched, if it is an ALU-related instruction, it can be executed by the ALU as illustrated in Figure 3.6 in a single cycle.

On the other hand, the SRAM memory access takes two cycles, as illustrated in Figure 3.7. This is because the SRAM access uses a pointer register for the SRAM address. The pointer register is one of the pointer registers (X, Y, or Z register pairs). The first clock cycle is needed to access the register file and to operate upon the pointer register (the SRAM access instructions allow pre/post-address increment operation on the pointer register). At the end of the first clock cycle, the ALU performs this calculation, and then this address is used to access the SRAM location and to write into it (or read from it into the destination register), as illustrated in Figure 3.7.

3.6   I/O Memory

The I/O memory is the gateway to all the peripheral components of the AVR processor. It is implemented as SRAM and can be accessed in two ways: as SRAM as well as I/O registers. As SRAM, the addresses are beyond $20 to $5F and as I/O registers, the addresses start at $00 to $3F.

We will look at the I/O registers as registers rather than as SRAM. We will look at most I/O registers and the function of these registers. However, for a specific chip, it is advisable to refer to individual data sheets for up-to-date and accurate information. The discussion here refers to most common registers and their functions.

An important point to note here is regarding accessing the various I/O registers. To access the I/O registers, the AVR offers IN and OUT instructions. These instructions can

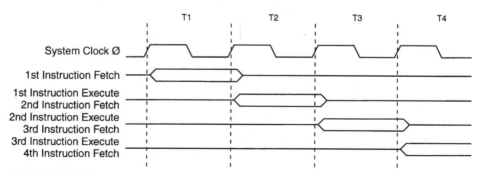

FIGURE 3.5 Instruction fetch/decode and instruction execution.

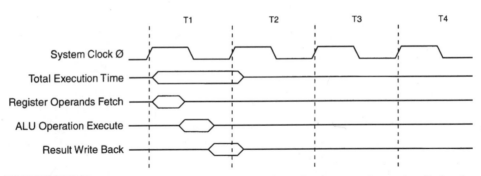

FIGURE 3.6 ALU execution consisting of register fetch, execute, and write back.

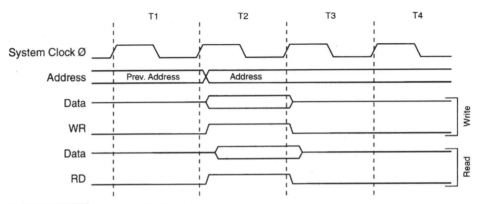

FIGURE 3.7 On-chip SRAM data access cycles.

access all the I/O registers from $00 to $3F. Besides IN and OUT, the AVR also supports bit addressing on some of the registers, namely from $00 to $1F. With the help of the bit instructions SBI and CBI, any bit in any of the registers ($00 to $1F) can be set on reset. This is a time-saving method compared to reading the register, changing the bit, and writing the value back to the register. For the rest of the registers, one has to use the other method, which takes about three times more clock cycles.

3.6.1 SREG: STATUS REGISTER

The STATUS register contains 8-flag bits that indicate the current state of the processor. All these bits are cleared (i.e., at logic "0") at reset and can be read or written to by the program. The I/O address of the STATUS register is $3F (memory address is $5F). (Figure 3.8)

The various flags of the STATUS register and their functions are:

1. Bit7-I: Global Interrupt Enable. Setting this bit enables all the interrupts. Resetting this disables all interrupts.
2. Bit6-T: Bit Copy Storage. Used with BLD (bit load) and BST (bit store) instruction for loading and storing bits from one register to another.

	7	6	5	4	3	2	1	0
I/O Address = $3F	I	T	H	S	V	N	Z	C
Initial Value	0	0	0	0	0	0	0	0

FIGURE 3.8 The processor STATUS register.

3. Bit5:H. Half Carry Flag. Indicates half carry in some arithmetic instructions.
4. Bit4:S. Sign Flag. This bit is the exclusive OR between the negative flag N and the Overflow flag V.
5. Bit3:V. Two's Complement Overflow Flag.
6. Bit2:N. Negative Flag.
7. Bit1:Z. Zero Flag. Indicates a zero result after an arithmetic or logical operation.
8. Bit0:C. Carry Flag. Indicates a carry in arithmetic or logical operation.

The STATUS register is not stored by the machine during an interrupt operation. The instruction in an interrupt routine can modify the STATUS flag bits, and so the user program must store and retrieve the STATUS register during an interrupt.

3.6.2 SP: STACK POINTER REGISTER

This register is 1 byte wide for processors that have up to 256 bytes of SRAM and is 2 bytes wide (called SPH and SPL) for those processors that have more than 256 bytes of SRAM. This register is used to point to the area in SRAM that is the top of the stack. The stack is used to store return addresses by the processor during an interrupt and subroutine call. Since the SP is initialized to $00 (or $0000 for a 2-byte SP) at reset, the user program must initialize the SP appropriately, as the SRAM starting address is not $00. SRAM starting address is $60. The stack grows down in memory address—i.e., pushing a value on stack results in the SP getting decremented. Popping a value out of stack increments the SP.

3.6.3 GIMSK: GENERAL INTERRUPT MASK REGISTER

The GIMSK register is used to enable and disable individual external interrupts by setting and resetting the concerned bit respectively. However, the interrupt to be actually serviced, the I bit in the STATUS register (SREG), must also be set to "1." (Figure 3.9.)

3.6.4 GIFR: GENERAL INTERRUPT FLAG REGISTER

The bits in GIFR indicate if an interrupt has occurred. If an external interrupt occurs, the corresponding INT flag in GIFR is set to "1." If the interrupt gets serviced (which happens if the I bit and the corresponding INT bit in GIMSK register is "1"), then the flag is reset. The flag can also be reset by writing a logical "1" to it. (Figure 3.10.)

	7	6	5	4	3	2	1	0
I/O Address = $3B	INT1	INT0						
Initial Value	0	0	0	0	0	0	0	0

FIGURE 3.9 The general interrupt mask register.

	7	6	5	4	3	2	1	0
I/O Address = $3A	INTF1	INTF0						
Initial Value	0	0	0	0	0	0	0	0

FIGURE 3.10 The general interrupt flag register.

3.6.5 MCUCR: MCU GENERAL CONTROL REGISTER

The bits in MCUCR allow general processor control. This includes external SRAM access enable/disable, sleep mode, and external interrupt sense control. (Figure 3.11.)

1. Bit7:SRE. External SRAM Enable. Setting this bit to "1" allows external SRAM access on processors that have the capability. PortA becomes AD0-7, PortC becomes A8-15, and WR* and RD* signals are activated on PortD as alternate pin functions. When this bit is "0," the ports function as normal ports and external SRAM access is disabled.
2. Bit6:SRW. External SRAM Access Wait State Bit. When this bit is "1," an extra wait state is inserted in the SRAM access cycle. Thus the SRAM is accessed in 4 cycles. When this bit is "0," the SRAM is accessed in 3 cycles.
3. Bit5:SE. Sleep Enable. Setting this bit to "1" enables the processor to go in one of the sleep modes. After setting this bit to "1," the program must execute the SLEEP instruction.
4. Bit4:SM. Sleep Mode. A "1" in this bit puts the processor in idle mode. A "0" means power down mode.
5. Bit3, 2:ISC11, ISC10. Interrupt sense control bit for INT1 as per Table 3.2.
6. Bit1, 0:ISC01, ISC00. Interrupt sense control bit for INT0 as per Table 3.3.

3.6.6 MCUSR: MCU STATUS REGISTER

The MCU status register provides information about the source of reset. (Figure 3.12.) The MCUSR contains 2 bits which indicate the source for the reset as per Table 3.4.

I/O Address = $35	7	6	5	4	3	2	1	0
	SRE	SRW	SE	SM	ISC11	ISC10	ISC01	ISC00
Initial Value	0	0	0	0	0	0	0	0

FIGURE 3.11 The MCU general control register.

TABLE 3-2 INTERRUPT1 SENSE CONTROL

ISC11	ISC10	DESCRIPTION
0	0	Low level on INT1 generates interrupt
0	1	Reserved
1	0	Falling edge on INT1 generates interrupt
1	1	Rising edge on INT1 generates interrupt

TABLE 3-3 INTERRUPT0 SENSE CONTROL

ISC01	ISC00	DESCRIPTION
0	0	Low level on INT0 pin generates interrupt
0	1	Reserved
1	0	Falling edge on INT0 pin generates interrupt
1	1	Rising edge on INT0 pin generates interrupt

I/O Address = $34	7	6	5	4	3	2	1	0
							EXTRF	PORF
Initial Value	0	0	0	0	0	0	0	0

FIGURE 3.12 The MCU status register.

3.6.7 TCCR0: TIMER/COUNTER0 CONTROL REGISTER

The Timer/Counter0 Control register is used to control the operation of the processor Timer/Counter0. This is a simple timer that counts up from the loaded count. The count is incremented at each clock signal at its input. The clock signal can be selected from one of seven sources as illustrated in Table 3.5. The eighth option allows the counter/timer to be stopped.

3.6.8 TCNT0: TIMER/COUNTER0 REGISTER

This is the actual timer/counter register. A value loaded in this register is used as the starting value, and the timer increments this value at each of its clock signals if the

TABLE 3-4   PORF AND EXTRF VALUES AFTER RESET. X MEANS UNDEFINED AND Y MEANS UNCHANGED

EXTRF	PORF	DESCRIPTION
X	1	Power-on reset
1	Y	External reset
Y	Y	Watchdog reset

TABLE 3-5   CLOCK0 PRESCALE SELECTION

CS02	CS01	CS00	DESCRIPTION
0	0	0	Stop the timer/counter0
0	0	1	CK
0	1	0	CK/8
0	1	1	CK/64
1	0	0	CK/256
1	0	1	CK/1024
1	1	0	External pin T0, falling edge
1	1	1	External pin T0, rising edge

counter/timer is enabled through the TCCR0 register. After the Timer/Counter0 overflows, it resets to $00 and continues counting up for each Timer/Counter0 clock signal. (Figures 3.13 and 3.14.)

3.6.9   TCCR1A: TIMER/COUNTER1 CONTROL REGISTER A

TCCR1A is a control register for Timer/Counter1. The signals for this registers are:

1. Bits 7,6:COM1A1, COM1A0: Compare Output Mode1, bits 1 and 0. The COM1A1 and COM1A0 control bits determine any output pin action following a compare match in Timer/Counter1. Any output pin actions affect pin OC1-Output Compare pin1. This is an alternative function to the I/O port, and the corresponding direction control bit must be set to "1" to control an output pin. For devices with 2 compare functions, bits 5 and 4 of the control register have similar functions to bits 7 and 6. The control configuration is illustrated in Table 3.6.
2. Bits 1,0-PWM11, PWM10: Pulse Width Modulator Select Bits. These bits select PWM operation of Timer/Counter1 as specified in Table 3.7. (Also see Figure 3.15.)

	7	6	5	4	3	2	1	0
I/O Address = $33						CS02	CS01	CS00
Initial Value	0	0	0	0	0	0	0	0

FIGURE 3.13 The Timer/Counter0 control register.

	7	6	5	4	3	2	1	0
I/O Address = $32	MSB							LSB
Initial Value	0	0	0	0	0	0	0	0

FIGURE 3.14 The Timer/Counter0 register.

TABLE 3-6   COMPARE1 MODE SELECT. X IS A OR B

COM1X1	COM1X0	DESCRIPTION
0	0	Timer/counter1 disconnected from output pin OC1X.
0	1	Toggle OC1X output.
1	0	Clear OC1X output to "0."
1	0	Set OC1X output to "1."

TABLE 3-7   PWM MODE SELECT

PWM11	PWM10	DESCRIPTION
0	0	PWM operation of timer/counter1 is disabled.
0	1	8-bit PWM.
1	0	9-bit PWM.
1	0	10-bit PWM.

3.6.10 TCCR1B: TIMER/COUNTER1 CONTROL REGISTER B

The bits of the TCCR1B register have the following:

1. Bit7:ICNC1:Input Capture1 Noise Canceler (4 CKs). When the ICNC1 bit is cleared to "0," the input capture trigger noise canceler function is disabled. The input capture is

	7	6	5	4	3	2	1	0
I/O Address = $2F	COM1A1	COM1A0	COM1B1	COM1B0			PWM11	PWM10
Initial Value	0	0	0	0	0	0	0	0

FIGURE 3.15 The Timer/Counter1 control RegisterA.

triggered at the first rising/falling edge sampled on the ICP input capture pin—as specified. When the ICNC1 bit is set to "1," four successive samples are measures on the ICP, input capture pin, and all samples must be high/low according to the input capture trigger specification in the ICES1 bit. The actual sampling frequency is the XTAL clock frequency.

2. Bit6:ICES1:Input Capture1 Edge Select. While the ICES1 bit is cleared to "0," the Timer/Counter1 contents are transferred to the Input Capture Register, ICR1, on the falling edge of the input capture pin, ICP. While the ICES1 bit is set to "1," the Timer/Counter1 contents are transferred to the Input Capture Register, ICR1, on the rising edge of the input capture pin, ICP.

3. Bit3:CTC1:Clear Timer/Counter1 on Compare Match. When the CTC1 control bit is set to "1," the Timer/Counter1 is reset to $0000 in the clock cycle after a compareA match. If the CTC1 control bit is cleared, Timer/Counter1 continues counting and is unaffected by a compare match. Since the compare match is detected in the CPU clock cycle following the match, this function will behave differently when a prescaling higher than 1 is used for the timer. When a prescaling of 1 is used, and the compareA register is set to C, the timer will count as follows if CTC1 is set:

```
... C-2  C-1  C  0  1 ...
```

When the prescaler is set to divide by 8, the timer will count like this:

```
... C-2, C-2, C-2, C-2, C-2, C-2, C-2, C-2  C-1, C-1, C-1, C-1,
    C-1, C-1, C-1, C-1  C, 0, 0, 0, 0, 0, 0, 0 ...
```

In PWM mode, this bit has no effect.

4. Bits 2,1,0-CS12, CS11, CS10:Clock Select1, bit 2,1 and 0. The ClockSelect1 bits 2,1 and 0 define the prescaling source of Timer/Counter1 similar to Timer/Counter0. (See Figure 3.16.)

3.6.11 TCNT1H, TCNT1L: TIMER/COUNTER1

This 16-bit register contains the prescaled value of the 16-bit Timer/Counter1. To ensure that both the high and low bytes are read and written simultaneously when the CPU accesses these registers, the access is performed using an 8-bit temporary register (TEMP). This temporary register is also used when accessing OCR1A and ICR1. If the main program and also interrupt routines perform access to registers using TEMP, interrupts must be disabled during access from the main program or interrupts if interrupts are re-enabled. (Figure 3.17.)

TCNT1 Timer/Counter1 Write: When the CPU writes to the high-byte TCNT1H, the written data is placed in the TEMP register. Next, when the CPU writes the low-byte TCNT1L, this byte of data is combined with the byte data in the TEMP register, and all 16 bits are written to the TCNT1 Timer/Counter1 register simultaneously. Consequently, the high-byte TCNT1H must be accessed first for a full 16-bit register write operation.

	7	6	5	4	3	2	1	0
I/O Address = $2E	ICNC1	ICES0			CTC1	CS12	CS11	CS10
Initial Value	0	0	0	0	0	0	0	0

FIGURE 3.16 The Timer/Counter1 control RegisterB.

	15	14	13	12	11	10	9	8	
I/O Address = $2D	MSB								TCNT1H
Initial Value	0	0	0	0	0	0	0	0	

	7	6	5	4	3	2	1	0	
I/O Address = $2C								LSB	TCNT1L
Initial Value	0	0	0	0	0	0	0	0	

FIGURE 3.17 The Timer/Counter1 register.

TCNT1 Timer/Counter1 Read: When the CPU reads the low-byte TCNT1L, the data of the low-byte TCNT1L is sent to the CPU and the data of the high-byte TCNT1H is placed in the TEMP register. When the CPU reads the data in the high-byte TCNT1H, the CPU receives the data in the TEMP register. Consequently, the low-byte TCNT1L must be accessed first for a full 16-bit register read operation. The Timer/Counter1 is realized as an up or up/down (in PWM mode) counter with read and write access. If Timer/Counter1 is written to and a clock source is selected, the Timer/Counter1 continues counting in the timer clock cycle after it is preset with the written value.

3.6.12 OCR1AH, OCR1AL: TIMER/COUNTER1 OUTPUT COMPARE REGISTERS

The output compare register is a 16-bit read/write register. The Timer/Counter1 Output Compare Register contains the data to be continuously compared with Timer/Counter1. Actions on compare matches are specified in the Timer/Counter1 Control and Status register.

Since the Output Compare Register, OCR1A, is a 16-bit register, a temporary register TEMP is used when OCR1A is written to ensure that both bytes are updated simultaneously. When the CPU writes the high byte, OCR1AH, the data is temporarily stored in the TEMP register. When the CPU writes the low byte, OCR1AL, the TEMP register is simultaneously written to OCR1AH. Consequently, the high-byte OCR1AH must be written first for a full 16-bit register write operation. The TEMP register is also used when accessing TCNT1 and ICR1. If the main program and interrupt routines also perform access to registers using TEMP, interrupts must be disabled during access from the main program or interrupts if interrupts are re-enabled.

3.6.13 OCR1BH, OCR1BL: TIMER/COUNTER1 OUTPUT COMPARE REGISTERS

The output compare registers are 16-bit read/write registers.

The Timer/Counter1 Output Compare Registers contain the data to be continuously compared with Timer/Counter1.

Actions on compare matches are specified in the Timer/Counter1 Control and Status register. A compare match only occurs if Timer/Counter1 counts to the OCR value. A software write that sets TCNT1 and OCR1A or OCR1B to the same value does not generate a compare match.

A compare match will set the compare interrupt flag in the CPU clock cycle following the compare event. Since the Output Compare Registers—OCR1A and OCR1B—are 16-bit registers, a temporary register TEMP is used when OCR1A/B are written to ensure that both bytes are updated simultaneously. When the CPU writes the high byte, OCR1AH or OCR1BH, the data is temporarily stored in the TEMP register. When the CPU writes the low byte, OCR1AL or OCR1BL, the TEMP register is simultaneously written to OCR1AH or OCR1BH. Consequently, the high byte OCR1AH or OCR1BH must be written first for a full 16-bit register write operation.

The TEMP register is also used when accessing TCNT1 and ICR1. If the main program and interrupt routines perform access to registers using TEMP, interrupts must be disabled during access from the main program and from interrupt routines if interrupts are allowed from within interrupt routines.

	15	14	13	12	11	10	9	8	
I/O Address = $2B	MSB								OCR1AH
Initial Value	0	0	0	0	0	0	0	0	

	7	6	5	4	3	2	1	0	
I/O Address = $2A								LSB	OCR1AL
Initial Value	0	0	0	0	0	0	0	0	

FIGURE 3.18 The Timer/Counter1 output compare RegisterA.

	15	14	13	12	11	10	9	8	
I/O Address = $29	MSB								OCR1BH
Initial Value	0	0	0	0	0	0	0	0	

	7	6	5	4	3	2	1	0	
I/O Address = $28								LSB	OCR1BL
Initial Value	0	0	0	0	0	0	0	0	

FIGURE 3.19 The Timer/Counter1 output compare RegisterB.

3.6.14 ICR1H, ICR1L: Timer/Counter1 Input Capture Registers

The input capture register is a 16-bit read-only register. When the rising or falling edge (according to the input capture edge setting, ICES1) of the signal at the input capture pin, ICP, is detected, the current value of the Timer/Counter1 is transferred to the Input Capture Register, ICR1. At the same time, the input capture flag, ICF1, is set to "1." Since the Input Capture Register, ICR1, is a 16-bit register, a temporary register TEMP is used when ICR1 is read to ensure that both bytes are read simultaneously. When the CPU reads the low-byte ICR1L, the data is sent to the CPU and the data of the high-byte ICR1H is placed in the TEMP register. When the CPU reads the data in the high-byte ICR1H, the CPU receives the data in the TEMP register. Consequently, the low-byte ICR1L must be accessed first for a full 16-bit register read operation. The TEMP register is also used when accessing TCNT1 and OCR1A. If the main program and also interrupt routines perform access to registers using TEMP, interrupts must be disabled during access from the main program or interrupts if interrupts are re-enabled. (Figure 3.20.)

3.6.15 WDTCR: WATCHDOG TIMER CONTROL REGISTER

1. Bit4:WDTOE. Watchdog Turn Off Enable. This bit is used in conjunction with the WDE bit. This bit is set to "1," when WDE is cleared to "0" to disable the watchdog timer. The processor clears this bit after four clock cycles.
2. Bit3:WDE. Watchdog Enable. When set to "1," the watchdog timer is enabled. To disable the watchdog, this bit is cleared to "0" and the WDTOE is set to "1." To disable the watchdog timer, the following procedure is employed: In a single operation, set WDTOE and WDE to "1." Clear WDE to "0" within next 4 clock cycles. This will then disable the watchdog timer.
3. Bit2-0:WDP2, WDP1, WDP0. Watchdog Timer Prescaler. These bits are used as in Table 3.8 to select the watchdog timer timeouts. (See Figure 3.21.)

3.6.16 EEAR: EEPROM ADDRESS REGISTER

The EEPROM Address register is two bytes wide for processors with more than 256 bytes of EEPROM and one byte wide for the rest.

	15	14	13	12	11	10	9	8	
I/O Address = $25	MSB								ICR1H
Initial Value	0	0	0	0	0	0	0	0	

	7	6	5	4	3	2	1	0	
I/O Address = $24								LSB	ICR1L
Initial Value	0	0	0	0	0	0	0	0	

FIGURE 3.20 The Timer/Counter1 input capture register.

TABLE 3-8 WATCHDOG TIMER PRESCALE SELECT

WDP2	WDP1	WDP0	WDT CYCLES	TYPICAL TIMEOUT AT 5 V
0	0	0	16 K	15 ms
0	0	1	32 K	30 ms
0	1	0	64 K	60 ms
0	1	1	128 K	120 ms
1	0	0	256 K	240 ms
1	0	1	512 K	490 ms
1	1	0	1024 K	970 ms
1	1	1	2048 K	1.9 s

	7	6	5	4	3	2	1	0
I/O Address = $21				WDTOE	WDE	WDP2	WDP1	WDP0
Initial Value	0	0	0	0	0	0	0	0

FIGURE 3.21 The watchdog timer control register.

3.6.17 EEDR: EEPROM DATA REGISTER

The EEPROM DATA register is used to read and write data from/to the EEPROM. The EEPROM is 8 bits wide.

3.6.18 EECR: EEPROM CONTROL REGISTER

The EECR is used to control data read and write operations to the EEPROM. (Figure 3.22.)

1. Bit2:EEMWE:EEPROM Master Write Enable. Setting EEMWE to "1" and then setting EEWE to "1" only will write data in the EEDR register to the EEPROM. If EEMWE is set to "1," the hardware clears this bit to "0" after 4 clock cycles.
2. Bit1:EEWE. EEPROM Write Enable. When set to "1" while EEMWE is also "1," the EEDR data is written to the EEPROM at the address specified by the EEPROM Address register. The EEWE bit remains "1" during the write cycle, which may take up to 2.5 ms at 5V. After this time has elapsed, the EEWE is cleared by hardware to "0." The sequence for writing data to the EEPROM is as follows:

Wait till EEWE is cleared to "0."
Write EEPROM address to EEAR.
Write EEPROM data to EEDR.
Set EEMWE to "1" and within four clock cycles set EEWE to "1."

This will write the data in EEDR to the EEPROM location whose address is in EEAR.

3. Bit0:EERE:EEPROM Read Enable. To read EEPROM data, load EEAR with the correct address, set EERE to "1," and then clear EERE to "0." This will get the data in EEDR. Before starting a read cycle, the program would poll the EEWE flag till EEWE is "0" to ensure that the any write cycle is not in progress.

I/O Address = $1C	7	6	5	4	3	2	1	0
						EEMWE	EEWE	EERE
Initial Value	0	0	0	0	0	0	0	0

FIGURE 3.22 The EEPROM control register.

3.6.19 PORTB: PORTB DATA REGISTER

PORTB register is a read/write register. It is initialized at reset to $00. When programmed as an output, then writing to PORTB will allow you to change the logic state at the PORTB pins.

3.6.20 DDRB: PORTB DATA DIRECTION REGISTER

This register is used to control the direction of each of the pins of the PORTB. Writing a "0" (which is also the reset value) in any bit of this register will make the corresponding POTB bit as input, and writing a "1" will make it an output bit.

3.6.21 PINB: INPUT PINS ON PORTB

This is a read-only port, and with this you can read the logic at the physical pin of PORTB. PINB is not a register, and reading PINB allows you to read the logical values on the pins of PORTB.

3.6.22 PORTD: PORTD DATA REGISTER

Same function as PORTB register.

3.6.23 DDRD: PORTD DATA DIRECTION REGISTER

Same function as DDRB register.

3.6.24 PIND: INPUT PINS ON PORTD

Same function as PINB port.

3.6.25 SPI I/O DATA REGISTER

This is the read/write register used for data transfer between the register file and the SPI shift register. Writing to this register initiates data transmission, and reading from it causes the shift register receive buffer to be read. More details in the SPI port section. (Figure 3.23.)

3.6.26 SPI STATUS REGISTER

1. Bit7:SPIF. SPI Interrupt Flag. When a SPI serial transfer is complete and the SPIE bit in SPCR is set to "1" and the global interrupts are enabled, then the SPIF flag is set to "1." SPIF is cleared to "0" by the processor when the corresponding interrupt is executed. Alternatively, the SPIF bit is cleared by reading the SPI status register when SPIF is "1" and then accessing the SPI data register.

	7	6	5	4	3	2	1	0
I/O Address = $0F	MSB							LSB
Initial Value	0	0	0	0	0	0	0	0

FIGURE 3.23 The SPI data register.

	7	6	5	4	3	2	1	0
I/O Address = $0E	SPIF	WCOL						
Initial Value	0	0	0	0	0	0	0	0

FIGURE 3.24 The SPI status register.

2. Bit6:WCOL. Write Collision Flag. This bit is set if the SPI Data Register (SPDR) is written during a data transfer. This bit is cleared, together with the SPIF, to "0" by first reading the SPI Status Register when WCOL is set to "1" and then accessing the SPI Data Register. (Figure 3.24.)

3.6.27 SPI CONTROL REGISTER

1. Bit7:SPIE. SPI Interrupt Enable. This bit causes an SPI interrupt to be generated if the SPIF bit in the SPSR register is set and the global interrupts are enabled.
2. Bit6:SPE. SPI Enable. When this bit is set to "1," the SPI is enabled.
3. Bit5:DORD. Data Order. When set to "1," LSB of the data word is transmitted first. When cleared to "0," the MSB of the data word is transmitted first.
4. Bit4:MSTR. Master/Slave Select. When set to "1," the SPI port is in master mode and when cleared to "0," it is a slave port. If SS* is configured as input and is driven low, then the MSTR will be cleared to "0" and SPIF in SPSR will be set. The user will have to set MSTR to "1" again to start as master.
5. Bit3:CPOL. Clock Polarity. When set to "1," the SCK is high when idle and when cleared to "0," SCK is low when idle.
6. Bit2:CPHA. Clock Phase. Determines the active phase of the clock.
7. Bit1-0:SPR1, SPR0. SPI Clock Rate Select. These bits determine the SCK clock rate when configured as master, as per Table 3.9. If the device is a slave, these bits have no effect on the SCK frequency. (Figure 3.25.)

3.6.28 UART I/O DATA REGISTER

The UART Data I/O registers are actually two separate registers, sharing the same physical address. When data is written to this address, it gets written to the data transmit register, and when reading from this address it is read from the data receive register. (Figure 3.26.)

3.6.29 UART STATUS REGISTER

The UART status register is used to monitor the status of the UART. The significant bits of the USR are:

TTABLE 3-9   SCK FREQUENCY. FCL IS THE PRO-
CESSOR OSCILLATOR FREQUENCY

SPR1	SPR0	SCK FREQUENCY
0	0	Fcl/4
0	1	Fcl/16
1	0	Fcl/64
1	1	Fcl/128

	7	6	5	4	3	2	1	0
I/O Address = $0D	SPIE	SPE	DORD	MSTR	CPOL	CPHA	SPR1	SPR0
Initial Value	0	0	0	0	0	0	0	0

FIGURE 3.25 The SPI control register.

	7	6	5	4	3	2	1	0
I/O Address = $0C	MSB							LSB
Initial Value	0	0	0	0	0	0	0	0

FIGURE 3.26 The UART I/O data register.

1. Bit7:RXC:UART Receive Complete. When this bit is set to "1," it indicates that the UART has received a data byte from the receiver shift register. RXC is cleared by reading the UDR.
2. Bit6:TXC:UART Transmit Complete. This bit is set to "1" when a complete data byte including the stop bit is shifted out from the transmit shift register and no new data is written to the UDR. TXC is cleared to "0" by hardware by executing the corresponding interrupt handler or by software by writing a "1" to the TXC bit.
3. Bit5:UDRE:UART Data Register Empty. This bit is set to "1," when the data written to the UDR is transferred to the transmit shift register. This bit indicates that the UDR is ready to receive a new byte.
4. Bit4:FE:Framing Error. This bit is set to "1," when the incoming stop bit is "0" (when it should be "1"). The FE is cleared when the incoming stop bit is "1."
5. Bit3:OR:Overrun Error. This bit is set to "1," when a valid data in the UDR is not read before a new data is shifted in the UDR from the UART receiver shift register. (Figure 3.27.)

3.6.30   UART CONTROL REGISTER

1. Bit7:RXCIE:RX Complete Interrupt Enable. This bit when set to "1" causes the Receive Complete Interrupt when the RXC bit in the USR is set to "1" and the global interrupts are enabled.

	7	6	5	4	3	2	1	0
I/O Address = $0B	RXC	TXC	UDRE	FE	OR			
Initial Value	0	0	1	0	0	0	0	0

FIGURE 3.27 The UART status register.

2. Bit6:TXCIE:TX Complete Interrupt Enable. This bit when set to "1" causes the Transmit Complete Interrupt when the TXC bit in the USR is set to "1" and the global interrupts are enabled.

3. Bit5:UDRIE:UART Data Register Empty Interrupt Enable. When this bit is set to "1" and the UDRE bit in the USR sets to "1," the UDRE data register empty interrupt will be executed provided the global interrupts are enabled.

4. Bit4:RXEN:Receiver Enable. When this bit is set to "1," the UART receiver is enabled.

5. Bit3:TXEN:Transmitter Enable. This bit when set to "1" enables the transmitter. When disabling the transmitter by writing a "0" to this bit, the transmitter is disabled but not before any character in the transmit shift register or the UDR transmit register is shifted out.

6. Bit2:CHR9:9-bit Characters. When this bit is set to "1," the transmitted and received characters are 9 bits long besides the start and the stop bit. The 9th bit can be used as an extra stop bit or parity bit.

7. Bit1:RXB8:Receive Data Bit 8. When CHR9 is set to "1," the RXB8 is the 9th bit of the received character.

8. Bit0:TXB8:Transmit Data Bit 8. When CHR9 is set to "1," the TXB8 is the 9th data bit in the character to be transmitted. (Figure 3.28.)

3.6.31 UART BAUD RATE REGISTER

The baud rate generator for the UART is a frequency divider which provides the time ticks for the data transmission and reception according to the following equation:

$$BaudRate = Fck/ (16 * (UBRR + 1))$$

Fck is the system clock frequency. UBRR is the contents of the UART Baud Rate Register. (Figure 3.29.)

3.6.32 ACSR: ANALOG COMPARATOR CONTROL AND STATUS REGISTER

The ACSR is used to control the comparator operation as well as to monitor the comparator output.

1. Bit7:ADC:Analog Comparator Disable. When set to "1," the power to the comparator is switched off.

2. Bit5:ACO:Analog Comparator Output. This is the output of the comparator.

3. Bit4:ACI:Analog Comparator Interrupt Flag. This bit is set to "1" when a comparator event has triggered a comparator interrupt mode defined by ACIS1 and ACIS0. The

I/O Address = $0A	7	6	5	4	3	2	1	0
	RXCIE	TXCIE	UDRIE	RXEN	TXEN	CHR9	RXB8	TXB8
Initial Value	0	0	0	0	0	0	1	0

FIGURE 3.28 The UART control register.

I/O Address = $09	7	6	5	4	3	2	1	0
	MSB							LSB
Initial Value	0	0	0	0	0	0	0	0

FIGURE 3.29 The UART baud rate register.

comparator interrupt is executed if the ACIE bit is set to "1" and the global interrupts are enabled.

4. Bit3:ACIE:Analog Comparator Interrupt Enable. When set to "1," the analog comparator interrupt is enabled. When reset to "0," the comparator interrupt is disabled.

5. Bit2:ACIC:Analog Comparator Input Capture Enable. When set to "1," the comparator output is connected to the input capture front-end circuit of the Timer1.

6. Bit1,0:ACIS1, ACIS0:Analog Comparator Interrupt Mode Select. The combinations of these bits selects the interrupt modes as illustrated in Table 3.10. Also see Figure 3.30.

3.7 The EEPROM

All AVR controllers have on-chip EEPROM. The amount of EEPROM varies from 64 bytes on the AT90S1200, Tiny10/12 to 4Kbytes on the Mega103. The EEPROM is accessed through the EEPROM access registers, namely: EEPROM Address Register (EEAR), EEPROM Data Register (EEDR), and the EEPROM Control Register (EECR).

For those devices with more than 256 bytes of EEPROM, the EEAR is actually two registers, EEARL and EEARH. The EEAR (either as a single register or as a double register) is used to set the address of the EEPROM to which data is to be written or from which the data is to be read. The EEAR is a read/write register, i.e., the register can be read to see what EEPROM address has been set.

The EEDR is the EEPROM data register and is a read/write register. When you want to write data to the EEPROM, you load the required data into the EEDR. When you want to read data from the EEPROM, after the reading process is over, you read the EEDR for the data.

The EECR has the necessary control bits for reading and writing the EEPROM. Writing to an EEPROM is not as simple as writing to SRAM, for example. The Write access time for the EEPROM on the AVR controllers is of the order or 2.5 to 4.0 ms, depending upon the supply voltage. The EEWE control bit in the EECR allows the user to detect when a previously requested data has been written to the EEPROM and whether a new byte can be written.

The following piece of code illustrates how a byte of data can be read from the EEPROM.

TABLE 3-10   ACIS1, ACIS0 SETTINGS

ACIS1	ACIS0	INTERRUPT MODE
0	0	Interrupt on output toggle.
0	1	Reserved.
1	0	Interrupt on falling output edge.
1	1	Interrupt on rising output edge.

	7	6	5	4	3	2	1	0
I/O Address = $08	ACD		ACO	ACI	ACIE	ACIC	ACIS1	ACIS0
Initial Value	0	0	0	0	0	0	0	0

FIGURE 3.30 The analog comparator control and status register.

```
;————-EEPROM Data Read Start————
eep_notrdy:
    sbic EECR,1        ;skip if EEWE clear
    rjmp eep_notrdy    ;Waits until EEPROM ready
read:
    out EEAR, ZL       ;output address
    sbi EECR, 0        ;set EERE (Read-strobe) low
    nop                ;mandatory 2 cycle delay
    nop
    in read_reg, EEDR  ;inputs data
;————-EEPROM Data Read End————--
```

The following piece of code illustrates how a byte of data can be written to the EE-PROM. To prevent any failure of data write to the EEPROM, it is important to ensure that the EEPROM write sequence of setting the EEWE bit and the EEMWE bit occurs without interruption; therefore global interrupts are disabled prior to the critical write sequence of setting the EEWE and the EEMWE bit, and after this the interrupts are enabled. However, this should only be done if interrupts are at all being used in the system. If the interrupts are not being used, there is no need to unnecessarily enable the interrupts.

```
;—————————EEPROM Data Write————————-
eep_notrdy:
    sbic EECR,1        ;skip if EEWE clear
    rjmp eep_notrdy    ;Waits until EEPROM ready
write:
    out EEAR, ZL       ;output address
    out EEDR, write_data
    cli                ;disable all interrupts
    sbi EECR, 1        ;set EEWE (Write-enable)
    sbi EECR, 2        ;set EEMWE (Master Write-enable)
    sei                ;enable all interrupts
;—————————EEPROM Data Write End————————
```

There have been many reports of EEPROM data corruption, mainly if the supply voltage is too low for the EEPROM to operate properly. According to Atmel, the solution to preventing data corruption for the on-chip EEPROM is much the same as preventing

EEPROM data corruption for off-chip EEPROM IC. Specifically, EEPROM data corruptions at EEPROM address $00 have been reported.

The recommended solutions for preventing EEPROM date corruption are as follows:

1. Use a brown-out detector (BOD) to detect periods of time when the supply voltage is low and assert reset during such time. It is better to reset the system than to proceed with the possibility of EEPROM data corruption.
2. Avoid writing to the EEPROM during periods of low supply voltage. This is easier said than done.
3. Do not use EEPROM at all! This is a rather extreme solution. However, keeping in mind the reports of EEPROM data corruption at address $00, it may be worthwhile to avoid using the EEPROM address $00.

3.8   The I/O Ports

All of the AVR controllers have some amount of I/O, which ranges from 3 bits on the AT90S2323 to 48 bits on Mega103. All the output bits of the AVR controllers can sink 20 mA of current, which makes it very suitable to drive LEDs directly without the need of external buffers.

All the I/O ports have three I/O addresses associated with them. The three addresses are required for configuring the individual bits as input or output; the other address is required to output data to those (or all) bits configured as output, and the third address is required to read data from those (or all) pins configured as input.

These ports are labeled as DDRx, PORTx, PINx for a given port x. The DDRx is the data direction register. Writing a "1" to a bit in the DDR makes the corresponding bit as output bit in portx. Thereafter, to output a "1" on the port bit, the corresponding bit can be set or reset using the CBI or SBI instruction or an OUT instruction.

```
;Using CBI and SBI to write to ports
SBI DDRB, 1        ;make bit 1 as output bit on PORTB
CBI PORTB, 1       ;make PORTB bit 1 as "0"
SBI PORTB, 1       ;make PORTB bit 1 as "1"
```

Another way to change port values is by using the IN and OUT instructions:

```
;Using OUT instruction to write to ports
LDI R18, 0b00000010
OUT DDRB, R18      ;make bit 1 as output bit on PORTB
LDI R18, 0b00000000
OUT PORTB, R18     ;make PORTB bit 1 as "0"
LDI R18, 0b00000010
OUT PORTB, R18     ;make PORTB bit 1 as "1"
```

Similarly, to read data at the input pin of a port, the PINx register is used. The PINx is directly connected to the pin of the port. The port pin can be provided with an internal pullup by writing a "1" to the port bit at the addresses PORTx. The value of this pullup resistor is between 30 Kohm and 150 Kohm. The corresponding value of the pullup current is between 160 μA and 33 μA.

Instead, if a "0" is written to the port bit at address PORTx, then the pullup is removed and the input pin is left floating in a high-impedance state.

```
;Using IN instruction to read from ports
LDI R18, 0b00000000
OUT DDRB, R18          ;make all bits as input bit on PORTB
LDI R18, 0b11111111
OUT PORTB, R18         ;Enable the pullup resistors on the PORTB
IN R18, PINB           ;read the pins on portB. R18 has the result.
```

See Figure 3.31.

3.9 The SRAM

The SRAM is available on most high-end processors. The amount varies from 128 bytes to 4 Kbytes. The SRAM is accessed using the many data access instructions either directly or indirectly using a pointer register. The SRAM is used for the stack also. The SRAM access time is two clock cycles, as illustrated in Figure 3.7.

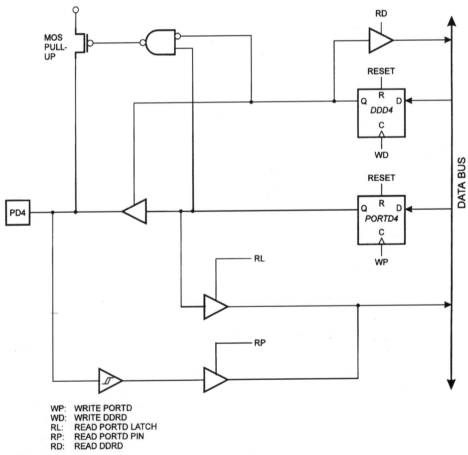

WP: WRITE PORTD
WD: WRITE DDRD
RL: READ PORTD LATCH
RP: READ PORTD PIN
RD: READ DDRD

FIGURE 3.31 Details of one of the port bits (PORTD4).

```
ldi r30, $60      ;init the pointer into SRAM to 60hex
clr r31           ;60hex is start of SRAM in 2313
                  ;now Z pointer is pointing to the SRAM
st z+, temp       ;store a value in register temp to SRAM
                  ;and increment the address
```

Similarly, to read the SRAM, the following code segment is used.

```
ldi r30, $60      ;init the pointer into SRAM to 60hex
clr r31           ;60hex is start of SRAM in 2313
                  ;now Z pointer is pointing to the SRAM
ld temp, z+       ;load a value in register temp from
                  ;SRAM and increment the address
```

3.9.1 INTERFACE TO EXTERNAL SRAM

On many larger AVR controllers, it is possible to connect external SRAM. This is illustrated in Figure 3.32. To enable the external SRAM access on PORTA and PORTC of the controllers as well as the ALE signal for address/data demultiplexing, the SRE bit (bit7) in MCUCR register is set to "1." The default access time for an external SRAM access is three clock cycles. This can be increased to four clock cycles by setting the SRW bit (bit6) in the MCUCR register. Figure 3.33 illustrates the normal three-cycle access, and Figure 3.34 illustrates the extended access cycle with an additional wait state.

3.10   The Timer

The timer in the AVR controller can function as a timer or a counter. As a timer, the internal clock signal or a derivative of that clock signal is used to clock the timer, while as a counter, an external signal on a port pin is used to clock the timer/counter. Figure 3.35 illustrates the

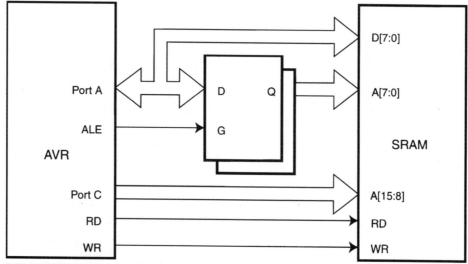

FIGURE 3.32 Connecting external SRAM to the AVR controllers.

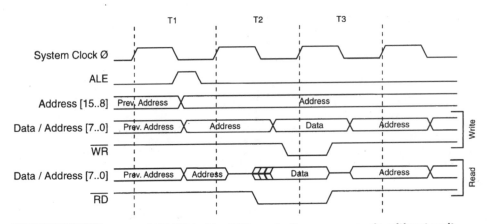

FIGURE 3.33 External SRAM to the AVR controller access cycle without wait states.

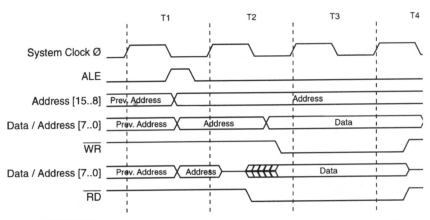

FIGURE 3.34 External SRAM to the AVR controller access cycle with additional wait states.

multiplexer which selects one of the many clock sources for the timer/counter. The prescaler for both the timer/counter0 and the timer/counter1 are illustrated in Figure 3.35.

The 8-Bit Timer/Counter0 block diagram is illustrated in Figure 3.36. The 8-bit Timer/Counter0 can select clock source from CK, prescaled CK, or an external pin. In addition, it can be stopped using the control bits in the Timer/Counter0 Control Register TCCR0.

The overflow status flag is found in the Timer/Counter Interrupt Flag Register TIFR. Control signals are found in the Timer/Counter0 Control Register TCCR0.

The interrupt enable/disable settings for Timer/Counter0 are found in the Timer/Counter Interrupt Mask Register TIMSK. When Timer/Counter0 is externally clocked, the external signal is synchronized with the oscillator frequency of the CPU. To ensure proper sampling of the external clock, the minimum time between two external clock transitions must be at least one internal CPU clock period. The external clock signal is sampled on the rising edge of the internal CPU clock. The 8-bit Timer/Counter0 features both a high-resolution and a high-accuracy usage with the lower prescaling opportunities. (Figure 3.37.)

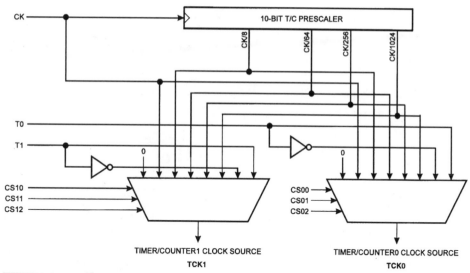

FIGURE 3.35 A clock prescaler for Timer0 as well as Timer1.

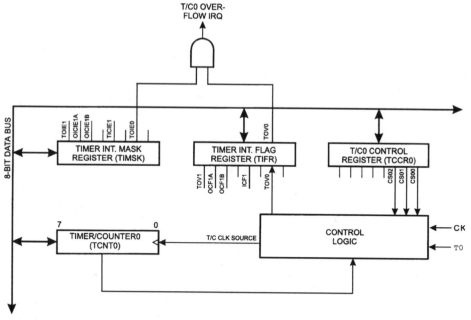

FIGURE 3.36 Timer/Counter0 block diagram.

3.11 The UART

Figure 3.38 illustrates the block diagram of the UART transmitter section. Data transmission is initiated by writing the data to be transmitted to the UART I/O Data Register, UDR. Data is transferred from UDR to the Transmit shift register when:

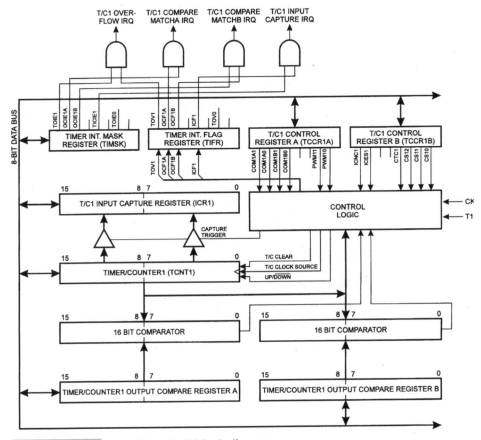

FIGURE 3.37 Timer/Counter1 block diagram.

A new character has been written to UDR after the stop bit from the previous character has been shifted out. The shift register is loaded immediately.

A new character has been written to UDR before the stop bit from the previous character has been shifted out. The shift register is loaded when the stop bit of the character currently being transmitted has been shifted out.

If the 10(11)-bit Transmitter shift register is empty, data is transferred from UDR to the shift register. At this time the UDRE (UART Data Register Empty) bit in the UART Status Register, USR, is set. When this bit is set to "1," the UART is ready to receive the next character. At the same time as the data is transferred from UDR to the 10(11)-bit shift register, bit 0 of the shift register is cleared (start bit) and bit 9 or 10 is set (stop bit). If a 9-bit data word is selected (the CHR9 bit in the UART Control Register, UCR is set), the TXB8 bit in UCR is transferred to bit 9 in the Transmit shift register. On the Baud Rate clock following the transfer operation to the shift register, the start bit is shifted out on the TXD pin. Then follows the data, LSB first. When the stop bit has been shifted out, the shift register is loaded if any new data has been written to the UDR during the transmission. During loading, UDRE is set to "1." If there is no new data in the

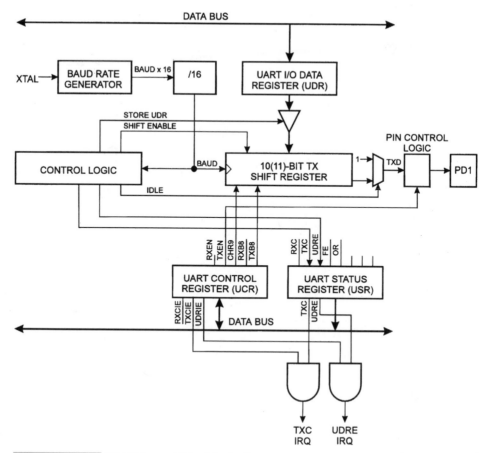

FIGURE 3.38 UART transmitter block diagram.

UDR register to send when the stop bit is shifted out, the UDRE flag will remain set until UDR is written again. When no new data has been written, and the stop bit has been present on TXD for one bit length, the TX Complete Flag, TXC, in USR is set to "1." The TXEN bit in UCR enables the UART transmitter when set to "1." When this bit is cleared to "0," the PD1 pin can be used for general I/O. When set to "1," the UART Transmitter will be connected to PD1, which is forced to be an output pin regardless of the setting of the bit 1 in DDRD.

Figure 3.39 illustrates the block diagram of the UART receiver section. The receiver front-end logic samples the signal on the RXD pin at a frequency 16 times the baud rate. While the line is idle, one single sample of logical zero will be interpreted as the falling edge of a start bit, and the start bit detection sequence is initiated.

Let sample 1 denote the first zero-sample. Following the 1 to 0-transition, the receiver samples the RXD pin at samples 8, 9, and 10. If two or more of these three samples are found to be logical ones, the start bit is rejected as a noise spike and the receiver starts looking for the next 1 to 0-transition. If, however, a valid start bit is detected, sampling of the data bits following the start bit is performed. These bits are also sampled at samples 8,

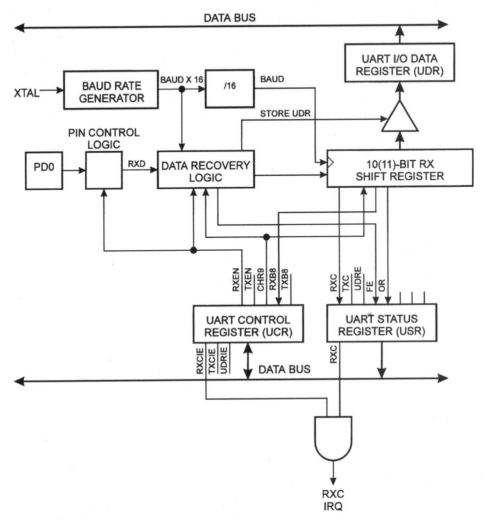

FIGURE 3.39 UART receiver block diagram.

9, and 10. The logical value found in at least two of the three samples is taken as the bit value. All bits are shifted into the transmitter shift register as they are sampled.

When the stop bit enters the receiver, the majority of the three samples must be one to accept the stop bit. If two or more samples are logical zeros, the Framing Error (FE) flag in the UART Status Register (USR) is set to "1." Before reading the UDR register, the user should always check the FE bit to detect framing errors. Whether or not a valid stop bit is detected at the end of a character reception cycle, the data is transferred to UDR and the RXC flag in USR is set. UDR is in fact two physically separate registers, one for transmitted data and one for received data. When UDR is read, the Receive Data register is accessed, and when UDR is written, the Transmit Data register is accessed. If 9-bit data word is selected (the CHR9 bit in the UART Control Register, UCR is set to "1"), the RXB8 bit in UCR is loaded with bit 9 in the Transmit shift register when data is transferred to UDR. If after having received a character, the UDR register has not been read since the last receive, the OverRun (OR) flag in UCR is set to "1." This means that the last data byte shifted into to shift register could not be transferred to UDR and has been lost. The OR bit is buffered and is updated when the valid data byte in UDR is read.

Thus the user should always check the OR bit after reading the UDR register in order to detect any overruns if the baud rate is high or CPU load is high. When the RXEN bit in the UCR register is cleared to "0," the receiver is disabled. This means that the PD0 pin can be used as a general I/O pin.

When RXEN is set to "1," the UART Receiver will be connected to PD0, which is forced to be an input pin regardless of the setting of the DDD0 bit in DDRD. When PD0 is forced to input by the UART, the PORTD0 bit can still be used to control the pull-up resistor on the pin. When the CHR9 bit in the UCR register is set, transmitted and received characters are 9 bits long, plus start and stop bits. The 9th data bit to be transmitted is the TXB8 bit in the UCR register. This bit must be set to the wanted value before a transmission is initiated by writing to the UDR register. The 9th data bit received is the RXB8 bit in the UCR register.

3.12   The Interrupt Structure

An interrupt is a flow control mechanism that is implemented on most controllers. In a processor system interacting with the outside world, many things are happening asynchronously, e.g., the user may have pressed a switch for some action to be taken, while a data byte on the serial port may have arrived. It would be quite impossible for the processor to keep track of all the things just by querying these devices for data. Instead, it would be better if these devices could "announce" arrival of data. This is what the interrupt mechanism does. The peripheral device could "interrupt" the execution of the main program, and the processor takes time out of the normal program execution to examine the source of the interrupt and to take necessary action. After the required action is taken, the interrupted program execution is resumed. The interrupt program is just like a subroutine, except that the execution of this interrupt subroutine is not anticipated by the processor to occur at a particular moment of time.

The AVR has a rich interrupt structure. Interrupt capability has been provided to most of the peripheral devices so that the main program need not poll these devices all the time.

The sequence of events when an interrupt occurs is as follows:

1. The peripheral device interrupts the processor.
2. Current instruction execution is completed.
3. The address of the next instruction is stored on the stack (either a hardware stack or a software stack).
4. Address of the ISR (interrupt subroutine) are loaded into the program counter.
5. The processor executes the ISR.
6. The ISR execution completion is indicated by the RETI instruction (return from interrupt).
7. The processor loads the program counter with the value stored on the stack and normal program execution resumes.

Since the interrupt can occur at any time, the processor status (flags, etc.) must be saved so that normal program execution can resume after the ISR is completed. The processor status is contained in the SREG register. The ISR must save the SREG before executing any other instruction, and before returning control to the main program, must restore the SREG register. This can be done in two ways: either the SREG is copied into another register, say R1, which must not be used for any other purpose, and before the ISR executes the RETI instruction, R1 is copied back into SREG. Another way to save the SREG is to save it on the stack (using the PUSH SREG instruction) and then before executing the RETI instruction, the SREG value is copied back from the stack (using the POP SREG instruction). This method is only possible for those processors that have a software stack. AT90S1200, for example, cannot use this method of saving the SREG register.

Figure 3.40 illustrates how the main program is interrupted. It is also possible to interrupt an ISR if another interrupt occurs and the global interrupt flag has been set to "1" within ISR for interrupt1 (using the SEI instruction). In that case, the ISR1 is interrupted and another ISR, ISR2, executes. ISR1 execution resumes after ISR2 finishes, and after ISR1 completes execution, the main program resumes execution.

Normally, after an interrupt occurs and is being serviced by the corresponding ISR, the global interrupts are disabled automatically (equivalent to executing the CLI instruction); however, it is possible to enable interrupts while an ISR is executing by executing the SEI instruction in the ISR. If another interrupt occurs during the time when an ISR is already operating, then it will be serviced by interrupting the original ISR.

The priority of interrupts is determined by the way the interrupt vectors are assigned. An interrupt vector at a lower address in the program memory has a higher priority. The priority of interrupt is used to decide which interrupt gets serviced first if more than one interrupt is pending at any moment of time. This situation can arise when global interrupts have been disabled in a system to allow some critical section of the program to execute. After the critical section is completed, the program enables the global interrupts. Now, during the time the critical section was being executed, two interrupts, an external Interrupt0 and UART Rx Complete interrupt occurred. Then, since the external Interrupt0 has a higher priority than the UART interrupts, the ISR corresponding to the external Interrupt0 will be executed, and after that, the ISR for the UART interrupt will execute.

The lowest program memory addresses are assigned for reset and interrupt vectors. For the AT90S2313, these vectors are corresponding program memory addresses as follows:

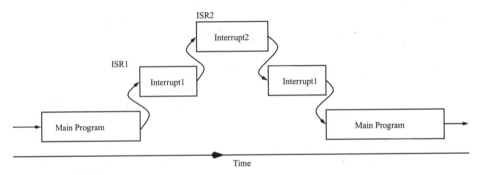

FIGURE 3.40 Nested interrupt execution.

ADDRESS	LABELS	CODE	COMMENTS
$000		rjmp RESET	; Reset Handler
$001		rjmp EXT_INT0	; IRQ0 Handler
$002		rjmp EXT_INT1	; IRQ1 Handler
$003		rjmp TIM_CAPT1	; Timer1 Capture Handler
$004		rjmp TIM_COMP1	; Timer1 Compare Handler
$005		rjmp TIM_OVF1	; Timer1 Overflow Handler
$006		rjmp TIM_OVF0	; Timer0 Overflow Handler
$007		rjmp UART_RXC	; UART RX Complete Handler
$008		rjmp UART_DRE	; UDR Empty Handler
$009		rjmp UART_TXC	; UART TX Complete Handler
$00a		rjmp ANA_COMP	; Analog Comparator Handler

A very important consideration while using interrupts is how fast a processor can respond to an interrupt. This is largely decided by the processor architecture. For the AVR controllers, the interrupt execution response for all the enabled AVR interrupts is four clock cycles minimum. Four clock cycles after the interrupt flag has been set, the program vector address for the actual interrupt handling routine is executed. During this four-clock-cycle period, the Program Counter (2 bytes) is pushed onto the Stack, and the Stack Pointer is decremented by 2. The vector is normally a relative jump to the interrupt routine, and this jump takes two clock cycles. If an interrupt occurs during execution of a multicycle instruction, this instruction is completed before the interrupt is served. A return from an interrupt handling routine takes four clock cycles. During these four clock cycles, the Program Counter (2 bytes) is popped back from the Stack, the Stack Pointer is incremented by 2, and the I flag in SREG is set. When the AVR exits from an interrupt, it will always return to the main program and execute one more instruction before any pending interrupt is served.

A sample program to understand the interrupt operation using Timer1 interrupt is available in the code directory as file intr_ex.asm. The hardware for this program is the same as in Figure 6.47 in Chapter 6 (and you don't need the EEPROM). Just connect the processor to the PC serial port and you can see the bit PORTB6 toggling at the rate of the Timer1 interrupt (in multiples of 10 ms).

3.13   The Internal Watchdog Timer

A watchdog timer is a controlled timer that is used as a wakeup device in case the software is lost in some infinite loop or in case of faulty program execution. The watchdog timer has an output that has the capability to reset the controller. Figure 3.41 illustrates the watchdog timer block diagram.

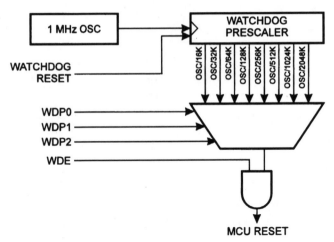

FIGURE 3.41 Watchdog timer block diagram.

The watchdog timer is clocked from a separate on-chip RC oscillator. By controlling the watchdog timer prescaler, the watchdog reset interval can be adjusted as illustrated in Table 3.8. The watchdog reset intervals are also power-supply dependent.

The watchdog reset instruction, WDR, resets the watchdog timer. Eight different clock cycle periods can be selected to determine the reset period. If the reset period expires without another watchdog reset, the AVR controller is reset and starts executing the program again from the reset vector. To prevent unintentional disabling of the watchdog, a special turn-off sequence must be followed when the watchdog is disabled, as illustrated in the description of the Watchdog Timer Control Register section.

3.14 Power-Down Modes of Operation

The AVR controller offers a variety of power-consumption-reducing schemes. To enter the sleep modes, the SE bit in MCUCR must be set (one) and a SLEEP instruction must be executed. If an enabled interrupt occurs while the MCU is in a sleep mode, the MCU awakes, executes the interrupt routine, and resumes execution from the instruction following SLEEP. The contents of the register file, SRAM, and I/O memory are unaltered. If a reset occurs during sleep mode, the MCU wakes up and executes from the reset vector.

When the SM bit is cleared (zero), the SLEEP instruction forces the MCU into the idle mode, stopping the CPU but allowing timer/counters, watchdog, and the interrupt system to continue operating. This enables the MCU to wake up from externally triggered interrupts as well as internal ones like timer overflow interrupt and watchdog reset. If wakeup from the analog comparator interrupt is not required, the analog comparator can be powered down by setting the ACD bit in the analog comparator control and status register ACSR. This will reduce power consumption in idle mode. When the MCU wakes up from idle mode, the CPU starts program execution immediately.

When the SM bit is set (one), the SLEEP instruction forces the MCU into the power-down mode. In this mode, the external oscillator is stopped, while the external interrupts and the watchdog (if enabled) continue operating. Only an external reset, a watchdog reset (if enabled), or an external level interrupt on INT0 or INT1 can wake up the MCU. Note that when a level-triggered interrupt is used for wake up from power down, the low level must be held for a time longer than the reset delay time-out period tTOUT. Otherwise, the device will not wake up. Figure 3.42 illustrates the current consumption by a Tiny22 processor during the active and power-down modes of operation.

3.15 Different Types of AVR Controllers

The AVR family offers many controllers with different peripheral resources, program memory, and packaging styles. Table 3.11 illustrates the available controllers and their features.

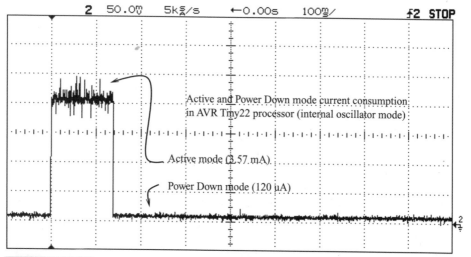

FIGURE 3.42 Current consumption by a Tiny22 processor in internal oscillator mode during the active and power down mode.

TABLE 3-11	AVR CONTROLLER SELECTION TABLE						
PART	PINS	SPEED	FLASH	EEPROM	RAM	UART	ADC
90S1200	20	16 MHz	1 K	64	0	No	No
90S2313	20	10 MHz	2 K	128	128	Yes	No
90S2323	8	10 MHz	2 K	128	128	No	No
90S2343	8	10 MHz	2 K	128	128	No	No
90S2333	28	10 MHz	2 K	128	128	Yes	Yes
90S4433	28	10 MHz	4 K	256	128	Yes	Yes

PART	PINS	SPEED	FLASH	EEPROM	RAM	UART	ADC
90S4414	40	10 MHz	4 K	256	256	Yes	No
90S8515	40	8 MHz	8 K	512	512	Yes	No
90S4434	40	10 MHz	4 K	256	256	Yes	Yes
Mega103	64	6 Mhz	128 K	4096	4096	Yes	Yes
Mega603	64	6 Mhz	64 K	2048	4096	Yes	Yes
Tiny10	8	10 MHz	1 K	64	0	No	No
Tiny12	8	10 MHz	1 K	64	0	No	No
Tiny13	8	10 MHz	2 K	128	128	No	No
Tiny22	8	10 MHz	2 K	128	128	No	No

4

THE AVR INSTRUCTION SET

The Instruction set of a processor or a controller is like the vocabulary of the processor. Each instruction controls some part of the processor and allows the programmer to manipulate data in the memory as well as input and output devices.

The instructions of the processor can be categorized in many different ways based on how the instructions access data and operate upon it. This is called the program and data addressing modes of the processor.

4.1   Program and Data Addressing Modes

The various AVR instructions can be categorized in about 10 different addressing modes. Each instruction has an opcode that indicates to the control logic of the processor what to do. The other part of the instruction is the operand, on which the opcode operates.

4.1.1   REGISTER DIRECT (SINGLE REGISTER)

The Register Direct instructions can operate on any of the 32 registers of the register file. It reads the contents of a register, operates on the contents of the register, and then stores the result of the operation back into the same register. Figure 4.1 illustrates the source and destination for these types of instructions.

The format of the instruction is: Mnemonic Destination Register.

Examples of these instructions are as follows. Rd is any register from the register file and is the destination (as well as the source) register for the operation.

COM Rd: 1's complement (invert all the bits) of the register Rd is stored back in register Rd.

INC Rd: Increments the contents of Rd by one.

DEC Rd: Decrements the contents of Rd by one.

TST Rd: Test for zero or negative contents of the Rd register.

CLR Rd: Loads $00 into the Rd register.

SER Rd: Loads $FF into the Rd register.

LSL Rd: Shifts the contents of register Rd one place to left. A "0" is shifted in bit position 0, and the contents of bit7 are copied to the Carry flag.

LSR Rd: Shifts the contents of register Rd one place to Right. A "0" is shifted in bit position 7, and the contents of bit0 are copied to the Carry flag.

ROL Rd: Rotate Rd register contents left through the carry. Carry flag goes to bit0, and bit7 goes into the carry.

ROR Rd: Rotate Rd register contents right through the carry. Carry flag goes to bit7, and bit0 goes into the carry.

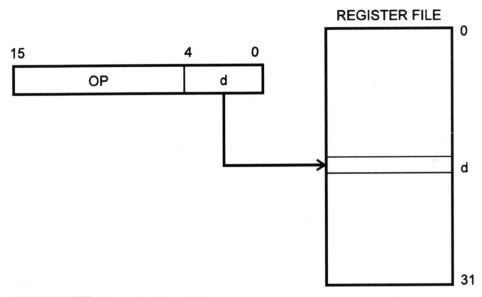

FIGURE 4-1 Direct single register access.

ASR Rd: Arithmetic Shift right the contents of the Rd, keeping the bit7 at the same place. This achieves a signed divide by two for each shift.

Swap Rd: Swap nibbles of the register Rd.

4.1.2 REGISTER DIRECT (TWO REGISTERS)

In these types of instructions, two registers are involved. The two registers are named as the source register, Rs, and the destination register, Rd. The instruction reads the two registers and operates on their contents and stores the result back in the destination register. Figure 4.2 illustrates the source and destination for these types of instructions.

Example instructions are: ADD Rd, Rs; SUB Rd, Rs; AND Rd, Rs; MOV Rd, Rs; OR Rd, Rs;

4.1.3 I/O DIRECT

These instructions are used to access the I/O space. The I/O registers can only be accessed using these instructions: In Rd, PORTADDRESS; Out PORTADDRESS, Rs.

Rd, Rs can be any of the 32 registers from the register file, and the I/O registers can be any register from the entire range of $00 to $3F (a total of 64 I/O registers). Figure 4.3 illustrates how such instructions operate.

4.1.4 DATA DIRECT

These are two word instructions. One of the words is the address of the data memory space. So a maximum of 64 Kbyte data memory can be accessed using these types of instructions.

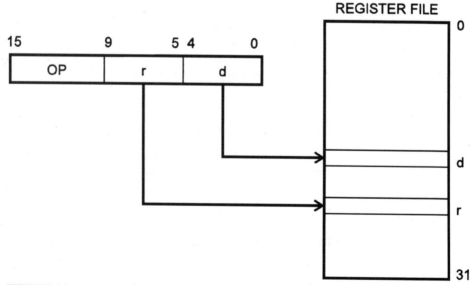

FIGURE 4-2 Direct double register access.

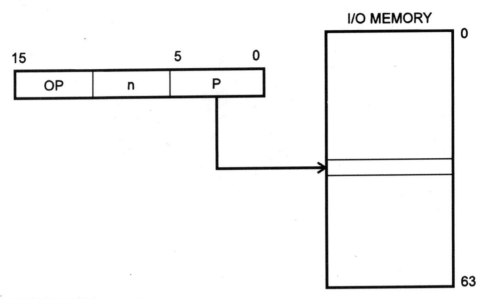

FIGURE 4-3 Direct I/O memory access.

The examples of these instructions are: LDS RD, K; K is a 16-bit address. STS K, Rs; Figure 4.4 illustrates how direct data instructions operate.

4.1.5 DATA INDIRECT

These are similar to the data direct type of instructions, except that these instructions are one word each, and a pointer register (X, Y, or Z) is used that has the base address of the data memory. To the base address in the pointer register, an offset can be added, as well as some increment/decrement operations on the pointer contents. Examples of these instructions are: LD Rd, X; X is the pointer register (register pair R26, R27); LD Rd, X+; Rd is the destination register and it is loaded with the contents of the data memory pointed to by the X register, and after the memory is accessed, the X register is incremented. ST X, Rs; ST X+, Rs; ST-Y, Rs; and so on. Figure 4.5 illustrates how one variant of indirect data instructions operate.

4.1.6 INDIRECT PROGRAM ADDRESSING

In these types of instructions, the Z register is used to point to the program memory. Up to 64 Kbytes of program memory can be accessed with the 16-bit Z register. Examples of these types of instructions are: IJMP and ICALL. Figure 4.6 illustrates how indirect program addressing instructions operate.

4.1.7 RELATIVE PROGRAM ADDRESSING

These instructions are of the type RJMP and RCALL, where an offset of +/−2K to the program counter is used. Figure 4.7 illustrates how relative program addressing instructions operate.

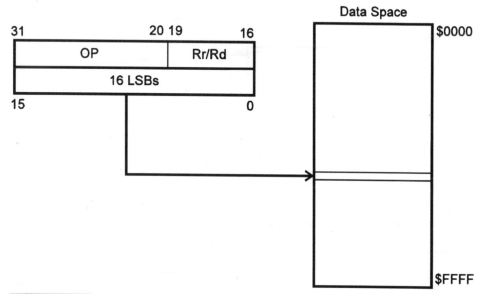

FIGURE 4-4 Direct data memory access.

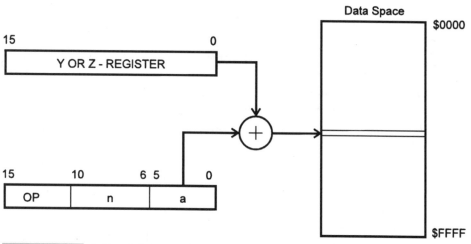

FIGURE 4-5 Indirect data memory access.

4.2   Arithmetic and Logic Instructions

1. ADD Rd, Rs; Add without Carry. Rd = Rd + Rs;
Flags affected: Z, C, N, V, S, H.
Clocks: 1.
Example: ADD R2, R3

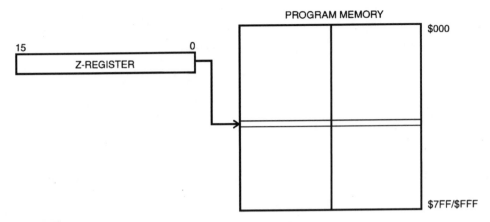

FIGURE 4-6 Indirect program memory instructions.

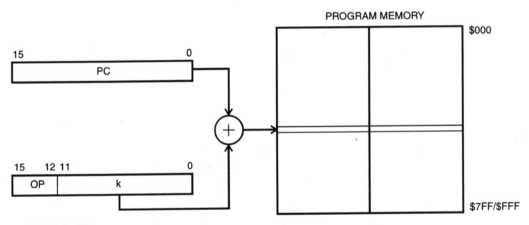

FIGURE 4-7 Indirect program memory instructions.

2. ADC Rd, Rs; Add with Carry. Rd = Rd + Rs + C;
Flags affected: Z, C, N, V, S, H.
Clocks: 1.
Example: ADC R5, R3

3. ADIW Rd, k; Add immediate constant to Rd:Rd+1. Rd+1: Rd = Rd + 1:Rd + k;
Flags affected: Z, C, N, V, S.
Clocks: 2.
Example: ADIW R26, 5

4. SUB Rd, Rs; Subtract without Carry. Rd = Rd − Rs;
Flags affected: Z, C, N, V, S, H.
Clocks: 1.
Example: SUB R1, R3

5. SUBI Rd, k; Subtract immediate without Carry. Rd = Rd − k;
    Flags affected: Z, C, N, V, S, H.
    Clocks:1.
    Example: SUBI R5, 10

6. SBCI Rd, k; Subtract immediate with Carry. Rd = Rd − k −C;
    Flags affected: Z, C, N, V, S, H.
    Clocks: 1.
    Example: SBCI R10, 2

7. SBC Rd, Rs; Subtract with Carry. Rd = Rd − Rs − C;
    Flags affected: Z, C, N, V, S, H.
    Clocks: 1.
    Example: SBC R5, R3

8. SBIW Rd, k; Subtract immediate constant from Rd:Rd + 1. Rd + 1: Rd
    = Rd + 1:Rd − k;
    Flags affected: Z, C, N, V, S.
    Clocks: 2.
    Example: SBIW R26, 5

9. AND Rd, Rs; Logical AND Rd and Rs. Rd = Rd & Rs.
    Flags affected: Z, N, V, S.
    Clocks: 1.
    Example: AND R26, R2

10. ANDI Rdx, k; Logical AND Rdx and k. Rdx = Rd & k. Rdx is between R16 and
    R31l; k is an 8-bit constant.
    Flags affected: Z, N, V, S.
    Clocks: 1.
    Example: ANDI R26, 5

11. OR Rd, Rs; Logical OR Rd with Rs. Rd = Rd | Rs.
    Flags affected: Z, N, V, S.
    Clocks: 1.
    Example: OR R1, R2

12. ORI Rdx, k; Logical OR Rdx with k. Rdx = Rd | k. Rdx is between R16 and R31; k
    is an 8-bit constant.
    Flags affected: Z, N, V, S.
    Clocks: 1.
    Example: ORI R16, 0b11110000

13. EOR Rd, Rs; Exclusive OR Rd with Rs. Rd = Rd Exor Rs.
    Flags affected: Z, N, V, S.
    Clocks: 1.
    Example: EOR R1, R2

14. COM Rd; One's complement Rd. Rd = $FF − Rd.
Flags affected: Z, C, N, V, S.
Clocks: 1.
Example: COM R1

15. NEG Rd; Two's complement Rd. Rd = $00 − Rd.
Flags affected: Z, C, N, V, S, H.
Clocks: 1.
Example: NEG R30

16. SBR Rdx, k; Set bit(s) in Rdx. Rdx = Rdx v k. Rdx is between R16 and R31.
Flags affected: Z, N, V, S.
Clocks: 1.
Example: SBR R30, 5

17. CBR Rdx, k; Clear bit(s) in Rdx. Rdx = Rdx & ($FF−k). Rdx is between R16 and R31.
Flags affected: Z, N, V, S.
Clocks: 1.
Example: CBR R28, 5

18. INC Rd; Increment contents of Rd. Rd = Rd + 1.
Flags affected: Z, N, V, S.
Clocks: 1.
Example: INC R3

19. DEC Rd; Decrement contents of Rd. Rd = Rd − 1.
Flags affected: Z, N, V, S.
Clocks: 1.
Example: DEC R2

20. TST Rd; Test Rd for zero or minus. Rd = Rd & Rd.
Flags affected: Z, N, V, S.
Clocks: 1.
Example: TST R2

21. MUL Rd, Rs; Unsigned multiplication of Rd and Rs. The result is stored in R1:R0.
R1: R0 = Rd × Rs.
Flags affected: Z, C.
Clocks: 2.
Example: MUL R3, R2

22. MULS Rd, Rs; Signed multiplication of Rd and Rs. The result is stored in R1: R0.
R1: R0 = Rd × Rs.
Flags affected: Z, C.
Clocks: 2.
Example: MULS R4, R5

23. MULSU Rd, Rs; Signed multiplication of Rd (signed) and Rs (unsigned). The result is stored in R1: R0. R1: R0 = Rd × Rs.
Flags affected: Z, C.
Clocks: 2.
Example: MULSU R4, R5

24. FMUL Rd, Rs; Unsigned fractional multiplication of Rd(1.7 format) and Rs(1.7 format). The result is stored in R1: R0(1.15 format). R1: R0 = Rd × Rs.
Flags affected: Z, C.
Clocks: 2.
Example: FMUL R30, R31

25. FMULS Rd, Rs; Signed fractional multiplication of Rd (1.7 format) and Rs (1.7 format). The result is stored in R1:R0 (1.15 format). R1: R0 = Rd × Rs.
Flags affected: Z, C.
Clocks: 2.
Example: FMULS R4, R5

26. FMULSU Rd, Rs; Fractional Signed multiplication of Rd (signed, 1.7 format) and Rs (unsigned, 1.7 format). The result is stored in R1: R0 (signed 1.15 format). R1: R0 = Rd × Rs.
Flags affected: Z, C.
Clocks: 2.
Example: FMULSU R4, R5

27. CLR Rd; Clear Register Rd

28. SER Rd; Set Register Rd

4.3   Program Control Instructions

1. RJMP k; Relative jump to a location in program memory. K is +/− 2K addresses.
Flags affected: none.
Clocks: 2.
Example:
 RJMP OK
 NOT_OK: ADD R1, R5
 OK: INC R1

2. IJMP; Indirect jump to a location in program memory pointed by the Z register.
Flags affected: none.
Clocks: 2.
Example:
 ldi r30, add_low
 ldi r31, add_high
 IJMP

3. JMP k; Jump to a location in program memory.
Flags affected: none.
Clocks: 3.
Example:
```
 JMP go_far
 ;
 ;
 ;
 go_far:
```

4. 4. RCALL k; relative call to a subroutine.
Flags affected: none.
Clocks: 3/4.
Example: rcall my_subroutine

5. ICALL; Indirect call to a subroutine.
Flags affected: none.
Clocks: 3/4.
Example:
```
 ldi r30, farsub_low
 ldi r31, farsub_high
 icall
```

6. CALL k; call to a subroutine.
Flags affected: none.
Clocks: 4/5.
Example: call my_subroutine

7. RET; Subroutine return.
Flags affected: none.
Clocks: 4/5.
Example:
```
 ;
 ;
 POP SREG
 RET
```

8. RETI; Interrupt return.
Flags affected: I.
Clocks: 4/5.
Example:
```
 ;
 ;
 POP SREG
 RETI
```

9. CPSE Rd, Rs; Compare, Skip if Equal. If Rd = Rs, skip next instruction.
Flags affected: none.
Clocks: 1/2/3
Example:
  back_here: CPSE r3, r4
  rjmp back_here
  in r3, PINB

10. CP Rd, Rs; Compare. Rd − Rs.
Flags affected: Z, C, N, V, S, H.
Clocks: 1
Example:
  CP r18, r19
  breq some_place

11. CPC Rd, Rs; Compare with carry. Rd − Rs − C.
Flags affected: Z, C, N, V, S, H.
Clocks: 1
Example:
  CP r18, r20
  CPC r19, r21
  brne some_place

12. CPI Rdx, k; Compare with carry. Rdx − k. Rd is between r16 and r31.
Flags affected: Z, C, N, V, S, H.
Clocks: 1
Example:
  CPI r18, 0
  breq some_place

13. SBRC Rd, b; Skip if bit in register cleared. If [Rd(b) = 0], skip next instruction.
Flags affected: none.
Clocks: 1/2/3
Example:
  back_here: in r3, PINB
  SBRC r3, 0
  rjmp back_here
  in r4, PIND

14. SBRS Rd, b; Skip if bit in register set. If [Rd(b) = 1], skip next instruction.
lags affected: none.
Clocks: 1/2/3
Example:
  back_here: in r3, PINB
  SBRS r3, 0

rjmp back_here
in r4, PIND

15. SBIC A, b; Skip if bit in I/O register cleared. If [A(b) = 0],
skip next instruction.
Flags affected: none.
Clocks: 1/2/3
Example:
  back_here: SBIC PINB, 0
  rjmp back_here
  in r4, PIND

16. SBIC A, b; Skip if bit in I/O register set. If [A(b) = 1], skip next instruction.
Flags affected: none.
Clocks: 1/2/3
Example:
  back_here: SBIS PINB, 0
  rjmp back_here
  in r4, PIND

17. BRBS s, k; Branch if status flag set. conditional relative branch.
If flag s in the SREG is set, then branch k relative to the PC.
Flags affected: none.
Clocks: 1/2

18. BRBC s, k; Branch if status flag cleared. conditional relative branch.
If flag s in the SREG is cleared, then branch k relative to the PC.
Flags affected: none.
Clocks: 1/2

19. BREQ k; Branch if Equal. If Z = 1, then branch relative.
Flags affected: none.
Clocks: 1/2

20. BRNE k; Branch if Not Equal. If Z = 0, then branch relative.
Flags affected: none.
Clocks: 1/2

21. BRCS k; Branch if Carry set. If C = 1, then branch relative.
Flags affected: none.
Clocks: 1/2

22. BRCC k; Branch if Carry cleared. If C = 0, then branch relative.
Flags affected: none.
Clocks: 1/2

23. BRSH k; Branch if same or higher. If C = 0, then branch relative.
Flags affected: none.
Clocks: 1/2

24. BRLO k; Branch if lower. If C = 1, then branch relative.
Flags affected: none.
Clocks: 1/2

25. BRMI k; Branch if minus. If N = 1, then branch relative.
Flags affected: none.
Clocks: 1/2

26. BRPL k; Branch if plus. If N = 0, then branch relative.
Flags affected: none.
Clocks: 1/2

27. BRGE k; Branch if greater or equal, signed. If (N Exor V) = 0, then branch relative.
Flags affected: none.
Clocks: 1/2

28. BRLT k; Branch if less than, signed. If (N Exor V) = 1, then branch relative.
Flags affected: none.
Clocks: 1/2

29. BRHS k; Branch if Half carry flag set. If H = 1, then branch relative.
Flags affected: none.
Clocks: 1/2

30. BRHC k; Branch if Half carry flag cleared. If H = 0, then branch relative.
Flags affected: none.
Clocks: 1/2

31. BRTS k; Branch if T flag set. If T = 1, then branch relative.
Flags affected: none.
Clocks: 1/2

32. BRTC k; Branch if T flag cleared. If T = 0, then branch relative.
Flags affected: none.
Clocks: 1/2

33. BRVS k; Branch if Overflow flag set. If V = 1, then branch relative.
Flags affected: none.
Clocks: 1/2

34. BRVC k; Branch if Overflow flag cleared. If V = 0, then branch relative.
Flags affected: none.
Clocks: 1/2

35. BRIE k; Branch if Interrupts enabled. If I = 1, then branch relative.
Flags affected: none.
Clocks: 1/2

36. BRID k; Branch if Interrupt disabled. If I = 0, then branch relative.
Flags affected: none.
Clocks: 1/2

4.4   Data Transfer Instructions

1. MOV Rd, Rs; Copy register. Rd = Rs.
Flags affected: None.
Clocks: 1.
Example: MOV R2, R18

2. MOVW Rd, Rs; Copy register pair. Rd+1: Rd = Rs + 1: Rs.
Flags affected: None.
Clocks: 1.
Example: MOV R26, R30

3. LDI Rdx, k; Load Immediate. Rdx = k. Rdx is between r16 and r31.
Flags affected: None.
Clocks: 1.
Example: LDI R18, $53

4. LDS Rd, k; Load Immediate. Rd = data memory(k).
Flags affected: None.
Clocks: 2.
Example: LDS R1, $5300

5. LD Rd, X; Load Indirect. Rd = data memory(X). X is the pointer register pair R26: R27.
Flags affected: None.
Clocks: 2.
Example: LD R1, X

6. LD Rd, X+; Load Indirect. Rd = data memory(X), X = X +1. X is the pointer register pair R26: R27.
Flags affected: None.
Clocks: 2.
Example: LD R15, X+

7. LD Rd, $-$X; Load Indirect. X = X $-$1, Rd = data memory(X). X is the pointer register pair R26: R27.
Flags affected: None.
Clocks: 2.
Example: LD R15, $-$X

8. LD Rd, Y; Load Indirect. Rd = data memory (Y). Y is the pointer register pair R28: R29.
Flags affected: None.
Clocks: 2.
Example: LD R1, Y

9. LD Rd, Y+; Load Indirect. Rd = data memory (Y), Y = Y + 1. Y is the pointer register pair R28: R29.
Flags affected: None.
Clocks: 2.
Example: LD R15, Y+

10. LD Rd, $-$Y; Load Indirect. Y = Y $-$ 1, Rd = data memory(Y). Y is the pointer register pair R28: R29.
Flags affected: None.
Clocks: 2.
Example: LD R15, $-$Y

11. LDD Rd, Y+q; Load Indirect with displacement. Rd = data memory(Y + q). Y is the pointer register pair R28: R29.
Flags affected: None.
Clocks: 2.
Example: LD R15, Y+2

12. LD Rd, Z; Load Indirect. Rd = data memory(Z). Z is the pointer register pair R30: R31.
Flags affected: None.
Clocks: 2.
Example: LD R1, Z

13. LD Rd, Z+; Load Indirect. Rd = data memory(Z), Z = Z + 1; Z is the pointer register pair R30: R31.
Flags affected: None.
Clocks: 2.
Example: LD R15, Z+

14. LD Rd, $-$Z; Load Indirect. Z = Z $-$ 1, Rd = data memory(Z). Z is the pointer register pair R30: R31.
Flags affected: None.

Clocks: 2.
Example: LD R15, −Z

15. LDD Rd, Z+q; Load Indirect with displacement. Rd = data memory(Z + q). Z is the pointer register pair R30: R31.
Flags affected: None.
Clocks: 2.
Example: LD R15, Z+5

16. STS Rs, k; Store Immediate. data memory(k) = Rs.
Flags affected: None.
Clocks: 2.
Example: STS $5300, R1

17. ST X, Rs; Store Indirect. data memory(X) = Rs. X is the pointer register pair R26: R27.
Flags affected: None.
Clocks: 2.
Example: ST X, R1

18. ST X+, Rs; Store Indirect. data memory(X) = Rs, X = X +1. X is the pointer register pair R26: R27.
Flags affected: None.
Clocks: 2.
Example: ST X+, R15

19. ST −X, Rs; Store Indirect. X = X −1, data memory(X) = Rs. X is the pointer register pair R26: R27.
Flags affected: None.
Clocks: 2.
Example: ST −X, R11

20. ST Y, RD; Store Indirect. data memory(Y) = Rs. Y is the pointer register pair R28: R29.
Flags affected: None.
Clocks: 2.
Example: ST Y, R1

21. ST Y+, Rs; Store Indirect. data memory(Y) = Rs, Y = Y +1; Y is the pointer register pair R28: R29.
Flags affected: None.
Clocks: 2.
Example: ST Y+, R5

22. ST −Y, Rs; Store Indirect. Y = Y − 1, data memory(Y) = Rs. Y is the pointer register pair R28: R29.
Flags affected: None.

Clocks: 2.
Example: ST −Y, R5

23. STD Y + q, Rs; Store Indirect with displacement. data memory(Y + q) = Rs. Y is the pointer register pair R28: R29.
Flags affected: None.
Clocks: 2.
Example: STD Y+2, R8

24. ST Z, RD; Store Indirect. data memory(Z) = Rs. Z is the pointer register pair R30: R31.
Flags affected: None.
Clocks: 2.
Example: ST Z, R1

25. ST Z+, Rs; Store Indirect. data memory(Z) = Rs, Z = Z +1; Z is the pointer register pair R30: R31.
Flags affected: None.
Clocks: 2.
Example: ST Z+, R5

26. ST −Z, Rs; Store Indirect. Z = Z − 1, data memory(Z) = Rs. Z is the pointer register pair R30: R31.
Flags affected: None.
Clocks: 2.
Example: ST −Z, R5

27. STD Z+q, Rs; Store Indirect with displacement. data memory(Z+q) = Rs. Z is the pointer register pair R30: R31.
Flags affected: None.
Clocks: 2.
Example: STD Z+2, R8

28. LPM; Load Program Memory. R0 = Program Memory(Z). Z is the pointer register pair R30: R31.
Flags affected: None.
Clocks: 3.

29. LPM Rd, Z; Load Program Memory. Rd 5 Program Memory(Z). Z is the pointer register pair R30: R31.
Flags affected: None.
Clocks: 3.
Example: LPM R2, Z

30. LPM Rd, Z+; Load Program Memory. Rd = Program Memory(Z), Z = Z + 1. Z is the pointer register pair R30: R31.
Flags affected: None.

Clocks: 3.
Example: LPM R20, Z+

31. IN Rd, A; Input from Input Port. Rd = I/OPort(A),
Flags affected: None.
Clocks: 1.
Example: IN r1, PINB

32. OUT A, Rs; Output to output Port. I/OPort(A) = Rs,
Flags affected: None.
Clocks: 1.
Example: OUT PORTB, R16

33. PUSH Rs; Push register on STACK. STACK = Rs,
Flags affected: None.
Clocks: 2.
Example: PUSH r1

34. POP Rd; Pop into register from STACK. Rd = STACK,
Flags affected: None.
Clocks: 2.
Example: POP r1

4.5   Bit and Bit-test Instructions

1. LSL Rd; Logical Shift Left. Rd(n+1) = Rd(n); Rd(0) = 0; C = Rd(7).
Flags affected: Z, C, N, V, H.
Clocks: 1.
Example: LSL r1

2. LSR Rd; Logical Shift Right. Rd(n) = Rd(n+1); Rd(7) = 0; C = Rd(0).
Flags affected: Z, C, N, V.
Clocks: 1.
Example: LSR r10

3. ROL Rd; Rotate Left though Carry. Rd(0) = C; Rd(n+1) = Rd(n); C = Rd(7).
Flags affected: Z, C, N, V, H.
Clocks: 1.
Example: ROL r13

4. ROR Rd; Rotate Right through Carry. Rd(7) = C; Rd(n) = Rd(n+1); C = Rd(0).
Flags affected: Z, C, N, V.
Clocks: 1.
Example: ROR r10

5. ASR Rd; Arithmetic Shift Right. Rd(n) = Rd(n+1), for n = 0 to 6;
Flags affected: Z, C, N, V.
Clocks: 1.
Example: ASR r10

6. SWAP Rd; Swap Nibbles. exchange Rd(3,2,1,0) with Rd(7,6,5,4)
Flags affected: None.
Clocks: 1.
Example: SWAP r10

7. BSET s; Flag Set. SREG(s) = 1;
Flags affected: SREG(s).
Clocks: 1.
Example: BSET 7

8. BCLR s; Flag Reset. SREG(s) = 0;
Flags affected: SREG(s).
Clocks: 1.
Example: BCLR 7

9. SBI A, s; Set bit s in I/O register A. I/O(A,s) = 1;
Flags affected: none.
Clocks: 1.
Example: SBI PORTD, 7

10. CBI A, s; Clear bit s in I/O register A. I/O(A,s) = 0;
Flags affected: none.
Clocks: 1.
Example: CBI PORTD, 7

11. BST Rs, s; Bit Store from Rs to T flag. T = Rs(s);
Flags affected: T.
Clocks: 1.
Example: BST R1, 2

12. BLD Rd, s; Bit Store from T flag to Rd. Rd(s) = T;
Flags affected: none.
Clocks: 1.
Example: BLD R4, 2

13. SEC; Set Carry flag. C = 1;
Flags affected: C.
Clocks: 1.
Example: SEC

14. CLC; Clear Carry flag. C = 0;
Flags affected: C.

Clocks: 1.
Example: CLC

15. SEN; Set Negative flag. N = 1;
Flags affected: N.
Clocks: 1.
Example: SEN

16. CLN; Clear Negative flag. N = 0;
Flags affected: N.
Clocks: 1.
Example: CLN

17. SEZ; Set Zero flag. Z = 1;
Flags affected: Z.
Clocks: 1.
Example: SEZ

18. CLZ; Clear Zero flag. Z = 0;
Flags affected: Z.
Clocks: 1.
Example: CLZ

19. SEI; Set Interrupt flag. I = 1;
Flags affected: I.
Clocks: 1.
Example: SEI

20. CLI; Clear Interrupt flag. I = 0;
Flags affected: I.
Clocks: 1.
Example: CLI

21. SES; Set Signed Test flag. S = 1;
Flags affected: S.
Clocks: 1.
Example: SES

22. CLS; Clear Signed Test flag. S = 0;
Flags affected: S.
Clocks: 1.
Example: CLS

23. SEV; Set Two's Complement flag. V = 1;
Flags affected: V.
Clocks: 1.
Example: SEV

24. CLV; Clear Two's Complement flag. V = 0;
Flags affected: V.
Clocks: 1.
Example: CLV

25. SET; Set T flag. T = 1;
Flags affected: T.
Clocks: 1.
Example: SET

26. CLT; Clear T flag. T = 0;
Flags affected: T.
Clocks: 1.
Example: CLT

27. SEH; Set Half Carry flag. H = 1;
Flags affected: H.
Clocks: 1.
Example: SEH

28. CLH; Clear Half Carry flag. H = 0;
Flags affected: H.
Clocks: 1.
Example: CLH

29. NOP; No Operation;
Flags affected: none.
Clocks: 1.
Example: NOP

30. SLEEP; Sleep;
Flags affected: none.
Clocks: 1.
Example: SLEEP

31. WDR; Watchdog Reset;
Flags affected: none.
Clocks: 1.
Example: WDR

AVR HARDWARE DESIGN ISSUES

What does it take to get a simple AVR-based circuit up and running? Well, it would take a processor, voltage source, clock generation circuit, and a suitable reset circuit for reliable operation.

Figure 5.1 illustrates this simple configuration. The external clock generation circuit is optional, as many of the AVR processors have an internal RC oscillator that is used when feasible. Let us look at these issues in some detail.

5.1   Power Source

The power source for running a processor system is a critical component. No system would run without a power supply. There are various options that the designer may consider, depending upon the application. Broadly, the choice would be dictated by whether the system is portable and hence must use a battery source or whether it is for a desktop application, where an AC power line could be used.

Sometimes you may have access to an AC power line, but the battery operation may seem more convenient simply because it offers added portability and does not require a bulky transformer and associated rectifier, filter, and regulator components.

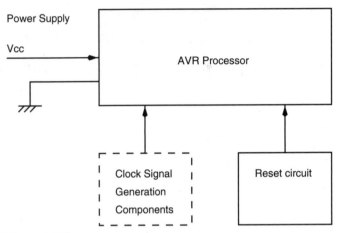

FIGURE 5.1 A minimum configuration AVR circuit.

5.1.1 BATTERY POWER

Batteries are of two types: primary batteries (nonrechargeable) and secondary batteries (these can be recharged). These are available in various shapes and sizes. While selecting a battery for your application, the following issues need to be considered:

1. Energy content or capacity. This is expressed in Ah (or mAh) (Ampere Hour or milliAmpere Hour). This is an important characteristic that indicates how long the battery can last before it discharges and becomes useless. For a given battery type, the capacity also dictates the battery size. A battery with a larger Ah rating will necessarily be bigger in volume than a similar battery with a smaller Ah rating.
2. Voltage. The voltage provided by the battery.
3. Storage. This indicates how the battery needs to be stored when not being used.
4. Shelf life. This indicates how long the battery will last before it discharges on its own. There is no point in buying a stock of batteries for the next 10 years if the shelf life of the batteries is, say, only 1 year.
5. Operating temperature. Batteries have notoriously poor temperature characteristics. This is because the batteries depend upon chemical reaction to produce power and the chemical reaction is temperature dependent. Batteries perform rather poorly at low temperatures.
6. Duty cycle. Some batteries perform for a longer period of actual usage time if they are used intermittently. The duty cycle of the battery indicates if the battery can be used continuously or not, without loss of performance.

Primary batteries Primary batteries are those that cannot be recharged. Once they lose energy, they have to be replaced. Primary batteries are of different types. Most common are the zinc chloride with carbon electrodes dry cells. The cell voltage is 1.5 V. These are the cheapest of all the primary batteries. Increasingly, alkaline cells also of 1.5 V are becoming popular. They have a higher capacity compared to the zinc chloride cells. Alkaline cells also have higher shelf life than the zinc chloride cells. Another type of primary battery is the lithium battery with a cell voltage of 3.0 V. These batteries are expensive compared to the zinc chloride and the alkaline batteries but have much higher energy density and shelf life of up to 10 years.

Secondary batteries   Secondary batteries have the advantage that they can be recharged after being discharged. Most popular of these batteries is the NiCd (nickel cadmium) and the lead acid batteries. The NiCd batteries have a cell voltage of 1.2 V, the so-called 9-V box type NiCd batteries are actually about 8.2 V. The lead-acid batteries have a cell voltage of 2.0 V. Lead-acid batteries of the so-called sealed variety are safe for use in portable instruments. Contrary to the lead-acid batteries used in cars, these do not pose any danger of leaking.

Lead-acid batteries have higher energy density than NiCd. The lead-acid batteries also have a relatively larger retention compared to the NiCd batteries.

Secondary batteries perform well if they are recharged regularly. If these batteries are discharged more than a certain minimum, their operational life reduces drastically.

Batteries are charged at a fraction of their Ah rating. Typically, lead-acid batteries are charged at a tenth of the Ah rating of the battery. NiCd batteries, on the other hand, are quite quirky. NiCd batteries are recommended to be initially charged at a tenth of the Ah rating and then switched over to trickle charging at a fraction (1/50) of the Ah rating.

5.1.2 MAIN OPERATING SUPPLY

Using AC wall supply is another alternative (and another being solar cells) to batteries. For embedded applications, a step-down transformer can either be integrated into the application or provided as a wall plug-in unit. Such units are very popular for small applications requiring a couple of watts of power.

Figure 5.2 illustrates the rectifier and filter unit that can be fed by a wall plug-in transformer. The rectifier could be built with discrete rectifier diodes (such as 1N4001), or even a complete rectifier unit be used. The rectifier should be suitably rated, keeping in mind the current requirements. If the power supply unit is to provide 500 mA of current, the diodes should be rated at at least 1A. The other rating of the diode to consider is the PIV (peak inverse voltage). This is the maximum peak reverse voltage that the diode can withstand before breaking down. An 1N4001 diode has a PIV of 50 V, and 1N4007 is rated to 1000 V.

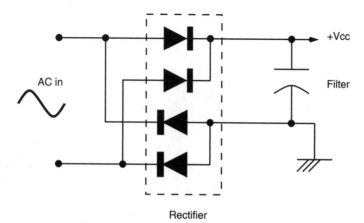

FIGURE 5.2 A rectifier and filter unit.

The peak rectified voltage that appears at the filter capacitor is 1.4 times the AC input voltage (AC input voltage is a RMS figure). A 10-V AC input will generate about 14 V DC voltage on the filter capacitor. The filter capacitor must be of sufficiently large capacity to provide sustained current. The filter capacitor must also be rated to handle the DC voltage. For a 14-V DC, at least a 25 V rating capacitor should be employed.

5.1.3 POWER FROM PORT SIGNAL LINES

Another source of power that is available most often if you are using a PC together with your application is the PC port signal lines, such as the parallel port and the serial RC232 port. If the PC is equipped with a USB port, it can be used to provide +5 V that is available on a USB connector.

I have used the RS-232 signal power in many low-power applications. The RS-232 signals can provide up to 10 mA each. The output signals of the RS-232 port that can be used to provide supply voltage are:

1. TxD: Output data signal from the RS-232 port. When idling, this signal is at −12 V and can be used to provide -ve voltage to the circuit.
2. DTR: Data Terminal Ready. Output signal used for communicating with a modem. This can be used to provide either +12 or −12 V supply. However, when the signal is loaded, the voltage drops. Usually, a micropower regulator or even a zener diode can be used to get a lower supply voltage for the AVR processor.
3. RTS: Request to Send. Output control signal used for communicating with an external device. Can be used in a fashion similar to the DTR signal.

The DTR and the RTS signals are controlled by the Modem Control Register (MCR) in a PC. The MCR is at an address offset 4 from the RS-232 base address. By writing $01 to the MCR, the DTR is set to +12 V and by writing $02, the RTS is set to +12 V, while writing $03 to the MCR, both the RTS and the DTR are set to +12. Table 5.1 illustrates all the combinations of voltages that you can get on the DTR and the RTS signal pins by writing the appropriate control word to the MCR.

The following C code (rspower.c) shows how to put both the DTR and the RTS signal to +12 V.

TABLE 5-1 MCR CONTROL VALUES FOR DTR AND RTS SIGNAL VOLTAGES		
MCR VALUE	**RTS VOLTAGE**	**DTR VOLTAGE**
0	−12 V	−12 V
1	−12 V	+12 V
2	+12 V	−12 V
3	+12 V	+12 V

```
/*rspower.c*/
/*Program segment to set DTR and RTS signal pins of the RS-232 port
(COM1) to +12V*/
#include <stdio.h>
#include <dos.h>
main()
{
int MCR;
MCR = peek(0x40, 0) +4; /*get the address of COM1 MCR regsiter*/
printf(``"nCOM1 MCR Address is: %x'', MCR);
outportb(MCR, 3); /*set MCR to assert DTR and RTS*/
}
```

Table 5.2 illustrates the voltage variation on the RTS signal pin when it is set to $+12$ V (nominal voltage) and then loaded with an AVR processor-based circuit. The power supply circuit for this arrangement is illustrated in Figure 5.3.

5.1.4 VOLTAGE REGULATORS

Voltage regulators are important to provide a stable voltage to the processor and the associated circuit, even though the input voltage may vary. Voltage regulators are broadly classified as linear or switching. The switching regulators are of two types: step up or step down. We shall look at some of the voltage regulators, especially the so-called micropower regulators.

It is very common to use the 78XX type of three-terminal regulator. This regulator is made by scores of companies and is available in many package options. To power the AVR processor, you would choose the 7805 regulator for $+5$-V output voltage. It can provide up to 1-A output current and can be fed a DC input voltage between 9 V to 20 V. You could also choose an LM317 three-terminal variable voltage regulator and adjust the output voltage with the help of two resistors between 1.25 V and above.

A voltage regulator is an active component, and when you use this to provide a stable output voltage, it also consumes some current. This current may run into tens of milliamperes and is called the quiescent or bias current. Micropower regulators are special voltage regulators that have extremely low quiescent current.

Micropower regulators The LP2950 and LP2951 are linear, micropower voltage regulators from National Semiconductor, with very low quiescent current (75 μA typ.) and very low dropout voltage (typ. 40 mV at light loads and 380 mV at 100 mA maximum current).

TABLE 5-2 RTS VOLTAGE VARIATION AS A FUNCTION OF LOAD

RTS PIN VOLTAGE	ZENER VOLTAGE	CURRENT BEING DRAWN	COMMENTS
11.18 V	—	0	No load voltage
8.12 V	5.15 V	5.94 mA	No processor in the circuit
7.78 V	4.62 V	6.32 mA	Processor in the circuit

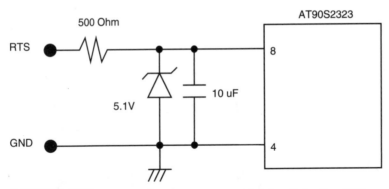

FIGURE 5.3 A super-simple power supply circuit for the AVR
processor powered by the RTS signal pin of the
RS-232 port and using a zener diode.

They are ideally suited for use in battery-powered systems. Furthermore, the quiescent current of the LP2950/LP2951 increases only slightly at higher dropout voltage. These are the most popular three-terminal micropower regulators. More information is available at this Web site:

www.national.com/pf/LP/LP2950.html

An LP2950 is also well suited for regulating voltage from the RS-232 signal pins instead of using the zener diode configuration as illustrated in Figure 5.3.

Table 5.3 illustrates a variety of micropower regulators. Some of these are switching regulators. The step-up type of switching regulators have an advantage that they can be used with batteries with lesser input voltage than required at the output.

5.2   Operating Clock Sources

Providing a clock source to the AVR processor is another important design process. The processor clock frequency determines the rate at which the programs will execute. On the AVR, most instructions execute in one clock cycle, some take two clocks, and for some high-end AVR processors, a few instructions also take four or five clock cycles.

The AVR processor clock can be operated with a variety of components. Many AVR processors are available with an internal RC oscillator. If the application does not demand timing accuracies, and if the nominal clock frequency of 1 MHz at 5 V is sufficient for the application, then this option can be utilized. Otherwise the internal clock generator with an external timing component such as a quartz crystal or a ceramic resonator can be used. If available, even an external TTL-level clock signal can be used to clock the processor.

5.2.1   USING A CRYSTAL CLOCK IC

A crystal clock is an integrated circuit available in 8- or 14-pin, DIP, or SMD package options. It contains all the components, active as well as the quartz crystal component, and just requires a supply voltage for operation. When using a crystal clock IC to drive the

TABLE 5-3   A SELECTION OF MICROPOWER VOLTAGE REGULATORS

COMPONENT	V OUT	QUIESCENT CURRENT	COMMENTS
MAX667	1.3 V to 16 V	20 μA	8 pins, linear regulator
MAX639	5.0 V	10 μA	8 pins, switching regulator
MAX630	V in to 18 V	70 μA	8 pins, switching step-up regulator
LP2950-5.0	5.0 V	40 μA	3 pin, linear regulator
LP2980-5.0	5.0 V	65 μA	5 pin, ultralow dropout linear regulator

clock input of the AVR processor, the clock signal is applied to the X1 input pin of the AVR processor.

The advantages of using a clock crystal IC are:

1. Industry standard package
2. CMOS/TTL output
3. 3.3-V operation available
4. Large fanout capability

More details are available at this manufacturer's Web site:

www.ndk.com/products/guide.htm

Figure 5.4 illustrates an 8-pin crystal clock and pinout.

5.2.2 USING A CERAMIC RESONATOR

For low-cost applications, a ceramic resonator is an attractive proposition. Ceramic resonators are three terminal components with the resonant components in a single package as illustrated in Figure 5.5.

Figure 5.6 illustrates how a ceramic resonator can be connected to the oscillator pins of the AVR processor. No other component is required for the oscillator.

The advantages of a ceramic resonator are:

1. Oscillation circuit does not require any additional capacitors
2. Wide frequency range of resonators are available
3. Small mechanical profile
4. No external adjustment required

Typical operating characteristics of a ceramic resonator are:

1. Frequency range: 2.0 to 60 MHz
2. Initial frequency tolerance: .5%
3. Frequency Stability: .3%
4. Frequency aging: .3% over 10 years' time

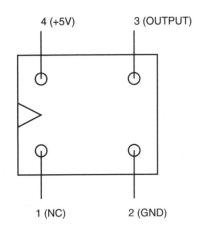

FIGURE 5.4 Crystal oscillator.

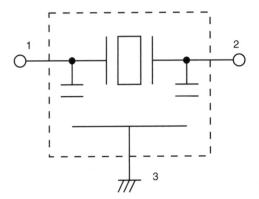

FIGURE 5.5 Ceramic resonator.

Ceramic resonators data sheets are available at the following Web sites:

www.token.com.tw/reson.htm

www.cirkit.co.uk/cirkit/PDFs/p16e9.pdf

5.2.3 USING A QUARTZ CRYSTAL

Using a quartz crystal is the most popular option. These are widely available and work without any problems. They require two additional capacitors and recommended values are between 22 pf and 33 pf to help start oscillations. The quartz crystals have a high Q (of the order of 10000 or more) and it takes some time for the oscillations to build up. This time is called start-up time and is of the order or 5 ms to 20 ms.

Figure 5.7 illustrates how to connect a quartz crystal to the oscillator pins of the AVR processor. If the clock signal is required for an external device, up to 1 HC type buffer or an inverter can be connected to the X2 pin of the processor.

When using an external crystal or resonator and in case of oscillation start-up problems, the oscillator start-up time should be investigated. Figure 5.8 illustrates the oscillator start-up time when power is applied to the AVR processor. The start-up time in this case is about 20 ms.

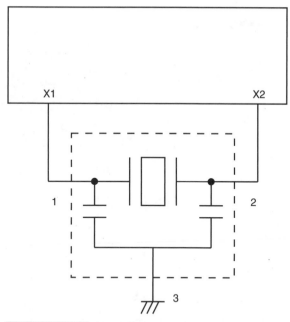

AVR Processor

FIGURE 5.6 Ceramic resonator connected to the oscillator pins of the AVR processor.

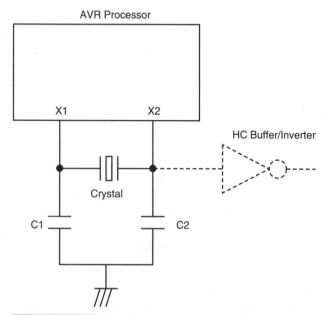

AVR Processor

FIGURE 5.7 A quartz crystal connection to the oscillator pins of the AVR processor.

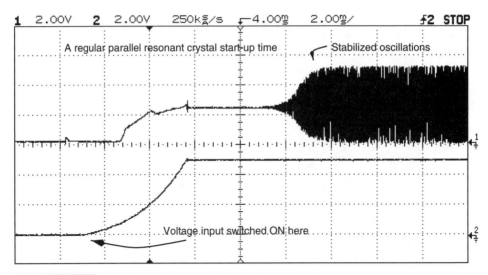

FIGURE 5.8 Oscillator start-up using a parallel resonant crystal after the supply input is applied to the processor.

5.2.4 USING A QUARTZ CLOCK CRYSTAL

For low-power operation, a general rule is to keep the operating frequency down. Typical operating frequency versus current graphs provided by the manufacturer substantiate this. However, such graphs supplied by the manufacturer usually plot frequency starting at about 1 MHz going up to 10–20 MHz or so. It seems that extrapolating the curve to lower frequency would result in a further-reduced operating current.

Lower frequency crystals, typically 32 kHz, are very commonly available for use in calendar clock circuits. It seems logical that using a 32-kHz crystal to generate the operating clock for an AVR circuit would reduce the operating current substantially.

I used the circuit illustrated in Figure 5.9 to test out this hypothesis. Figure 5.10 illustrates the start-up time for the 32-kHz crystal, which is about 2s. The oscillator showed start-up problems and I had to play with the C1, C2, and R1 values to get the circuit to oscillate. Table 5.4 illustrates my findings. However, more importantly, the current consumption is not reduced at all, as illustrated in Table 5.4.

Figure 5.11 illustrates an oscilloscope screen shot of the oscillations and the dynamic current consumption by the processor. The current consumption peaks during the times when the oscillations are in a transition phase.

It is concluded that the 32-kHz calendar clock crystal is not a viable component to be used with the AVR processor in terms of current savings.

5.2.5 USING INTERNAL RC CLOCK OSCILLATOR

The last option for clocking the AVR processor is to use the internal RC oscillator available on some of the processors (AT90S1200, 2343, Tiny22). The AT90S1200 is shipped with the internal RC oscillator disabled and can be reprogrammed with the help of a parallel programmer to enable the RCEN bit so as to select the RC oscillator. However, AT90S1200A can be used, which has the RC oscillator enabled.

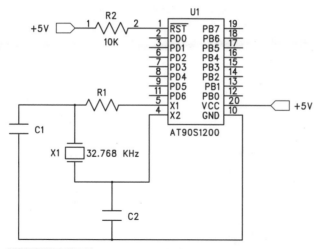

FIGURE 5.9 Circuit schematic for the 32-kHz clock crystal test circuit.

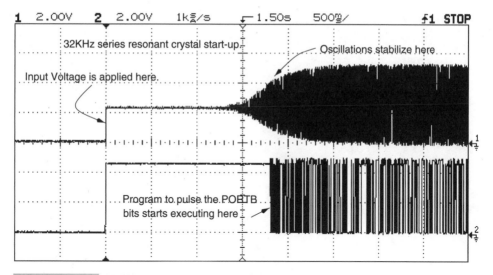

FIGURE 5.10 32-kHz oscillator start-up time.

The AT90S2343 and the Tiny22 are shipped with the RCEN bit enabled (i.e., at "0") and the internal RC oscillator can be used right away.

The RC oscillator has supply voltage dependence, and consequently this clocking option should only be used if the application does not require timing accuracy. I determined the frequency variation as a function of supply voltage with the help of a circuit illustrated in Figure 5.12. The supply voltage was varied between 2 V and 5.9 V and the corresponding pulse output was measured. The pulse output was the result of the program running on the 2343 processor, and the pulse frequency was related to the clock frequency as four clock cycles generating one pulse output cycle. By measuring the pulse frequency, the clock frequency was deduced and plotted, as illustrated in Figure 5.13 as well as in Table 5.5.

TABLE 5-4   32-KHZ OSCILLATOR START-UP TIMES AND CURRENT CONSUMPTION FOR VARIOUS CAPACITOR AND RESISTOR VALUES

R1	C1	C2	START-UP TIME (SECONDS)	I (DC) MA
82 K	33 pF	—	Doesn't oscillate	—
82 K	68 pF	—	8-10	5.83
82 K	68 pF	68 pF	8-10	5.5
120 K	68 pF	68 pF	3-5	5
120 K	68 pF	—	3-5	5
120 K	—	—	Doesn't oscillate	—
120 K	33 pF	—	10	5.7
390 K	68 pF	—	3-4	4
390 K	68 pF	33 pF	3	4
390 K	33 pF	33 pF	3	4
470 K	33 pF	33 pF	2-3	3.9
470 K	33 pF	—	2-3	3.9
470 K	68 pF	—	3-4	4
470 K	68 pF	68 pF	3	4
910 K	68 pF	68 pF	2-3	4.4
910 K	—	—	Doesn't oscillate	—
910 K	68 pF	—	4-5	4.4

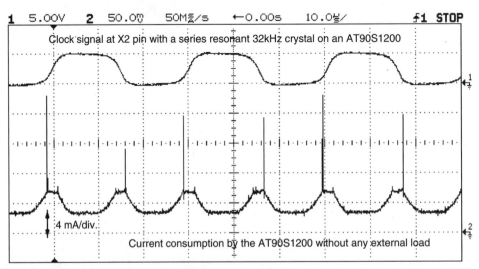

FIGURE 5.11 Current consumption by an AT90S1200 processor when operated with a 32-kHz clock crystal.

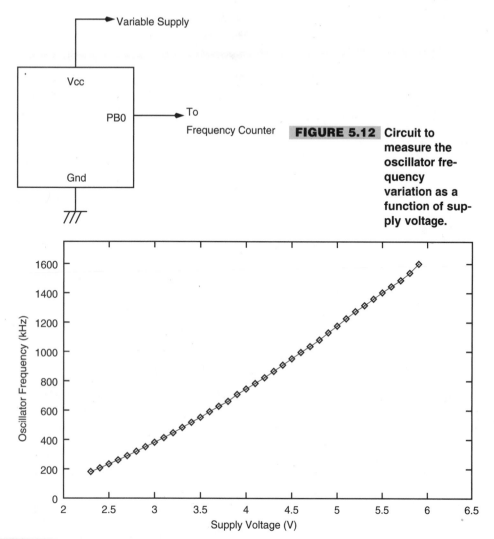

FIGURE 5.12 Circuit to measure the oscillator frequency variation as a function of supply voltage.

FIGURE 5.13 Variation of RC system clock frequency as a function of supply voltage.

The program rc_calib.asm in the code directory was used to measure the frequency of the output waveform generated on the PORTB pins. This frequency was then used to calculate the internal clock frequency of the processor.

5.3   Reset Circuit

CPUs require a reset pulse after the power supply has stabilized. The basic requirement is that a processor reset pulse should appear after the power supply has settled to a stable value. This is to initialize the internal registers and the control circuit. Usually, the processor has a power-on reset circuit as well as an external reset input circuit. The power-on reset circuit activates when the power supply voltage is below a certain threshold. After some

TABLE 5-5 VARIATION OF INTERNAL RC OSCILLATOR FREQUENCY WITH SUPPLY VOLTAGE

VCC(V)	FREQ (KHZ)	VCC(V)	FREQ (KHZ)	VCC(V)	FREQ (KHZ)
2.3	182	3.5	552	4.7	1036
2.4	207	3.6	591	4.8	1081
2.5	234	3.7	628	4.9	1130
2.6	261	3.8	662	5.0	1176
2.7	290	3.9	709	5.1	1226
2.8	320	4.0	746	5.2	1274
2.9	351	4.1	784	5.3	1315
3.0	381	4.2	823	5.4	1360
3.1	413	4.3	866	5.5	1403
3.2	446	4.4	909	5.6	1444
3.3	483	4.5	952	5.7	1486
3.4	518	4.6	995	5.8	1538
				5.9	1600

timeout period (called power-on reset period), the processor starts executing program memory code.

Figure 5.14 illustrates the simplest reset circuit using just a capacitor and a shunt switch for external reset. The reset pin of the AVR processor has an internal resistor of about 100-Kohm value between the reset pin and the Vcc supply voltage pin, and so any external resistor is not required. The capacitor is required to debounce the switch when it is pressed and released.

Figure 5.15 illustrates the signal on the reset pin of an AVR processor when the power is switched on. The processor is executing a program to generate pulses on the PORTB pins. The PORTB pins start pulsing after about 15 ms after the reset signal deactivates. This compares well with the datasheet specification for the timeout delay of 16 ms typically.

If the power supply voltage rises too slowly, then the rest can be extended by holding the external reset pin low for a longer time. This can be done by choosing either a large time constant RC circuit connected to the RESET pin or using an external reset.

If the power supply rises too slowly, the internal reset circuit may not be able to produce the proper system reset. An external reset generator circuit, often called a supervisory circuit, is useful for such needs. Some of the popular supervisory circuits are listed below. Some of them offer additional functions as well.

1. Dallas Semiconductor Corp:DS1236. 16-pin MicroManager chip. Active high and low reset, power fail signals.

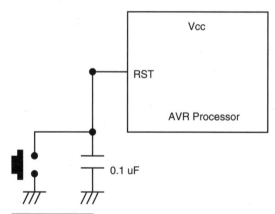

FIGURE 5.14 A simple reset circuit.

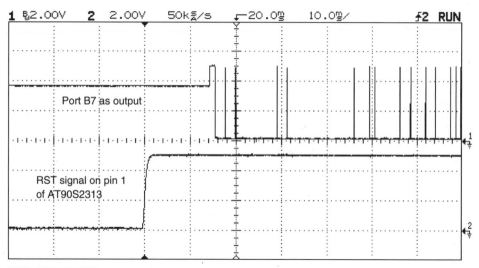

FIGURE 5.15 RST signal and the start of the program execution on an AVR processor.

2. Dallas Semiconductor Corp: DS1233. 3-pin reset generator. 350-ms pulse after Vcc supply voltage stabilizes. (Figure 5.16.)

3. Maxim Integrated Products Corp: MAX690

4. Maxim Integrated Products Corp: MAX809

5. Xicor Inc: X5045. Reset generator, 512 bytes of EEPROM, watchdog timer. 8-pin DIP.

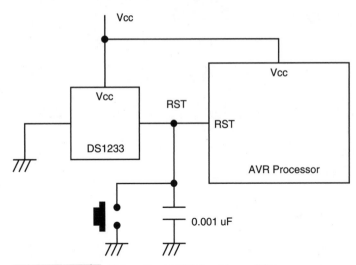

FIGURE 5.16 Using the DS1233 with an AVR processor.

HARDWARE AND SOFTWARE
INTERFACING WITH THE AVR

This chapter deals with actually putting the AVR processor to some use. It shows how to connect the AVR processor to many I/O devices such as switches, LEDs, displays, ADCs, DACs, motors, etc. To be able to do anything useful with a microcontroller, it needs the combination of appropriate hardware and suitable driver software. So the hardware and software for an embedded application for which the AVR processor could be used are tightly linked, and both of these aspects of a complete system design need to be considered together.

So let's get started and build our first supersimple circuit that will light up an LED. Trivial as it may seem, it nevertheless provides a lot of confidence to a beginner.

6.1   A Beginner's Circuit

If you are new to AVR processors, you probably want to build a simple circuit and run a program that does something. Nothing better than lighting up a LED. The circuit presented here and the code that runs on the processor does just that.

There are three aspects to this simple starting step:

1. Build the hardware on a general-purpose PCB.
2. Write the accompanying code and assemble it on a PC.
3. Program the AT90S1200 processor and plug it into your PCB.

You also need a +5-V power supply or at least three 1.5-V cells arranged in series to get about 4.5 V, which is suitable for running this circuit.

Figure 6.1 illustrates the circuit diagram. The circuit is not fancy at all. After you put the programmed chip into the socket and power the circuit, the LED should glow. Now press the switch connected to the reset pin and the LED should be turned off. Release the switch and the LED should glow again. This indicates that the program is running and it is the program that is lighting up the LED.

The circuit operates at 4 MHz using the external crystal. If you have a AT90S1200A part, then the oscillator components are not required and you can omit the crystal and the 22-pF capacitors. The AT90S1200A has the internal RC oscillator clock enabled, and the processor then runs at about 1 MHz at +5-V supply voltage. The clock speed is not critical in this particular case.

The following program is also available on the CD in the code directory as file led-light.asm.

```
;ledlight.asm
;A beginner's program
;lights up an LED on pin PORTB0
;LED is arranged to sink current into the PORTB0 pin
;assembled using Atmel's avrasm assembler.
;the following .inc file should be placed in the same directory as
;this assembly program
.include "1200def.inc"
.cseg
.org 0
```

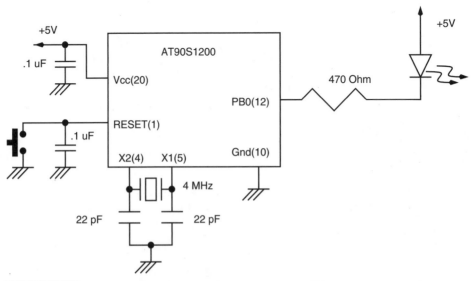

Figure 6.1 A simple introductory circuit to light an LED.

```
        rjmp    RESET              ;Reset Handle
        rjmp    RESET
        rjmp    RESET
RESET:  ldi r16, 0b11111111        ;load register r16 with all 1's
        out DDRB, r16              ;configure PORT B for all outputs
loopit: ldi r16, 0                 ;load register r16 with all 0's
        out PORTB, r16             ;output the contents of r16
                                   ;on PORTB
                                   ;Thus PORTB0 pin is at logic '0'
                                   ;as well as all the other PORTB
                                   ;pins. This enables the current
                                   ;through the LED to flow into
                                   ;PORTB0 pin and the LED lights up

        rjmp loopit
```

6.2   Lights and Switches

Now that we have built a simple beginner's circuit, let's add some input components to the circuit. The simplest input device is a switch. Figure 6.2 illustrates the circuit. The output devices, namely the LEDs, are connected to the PORTB pins, and the input devices, the switches, are connected to the PORTD pins. This keeps our code quite simple. Also, the LEDs and the switches are arranged in a logically symmetrical order and in our code, we map each switch to an LED. To keep code simple, let's map switch on PORTD0 pin to the LED on PORTB0 and so on.

What we want to do is to simply record the state of the switches and copy the state to the corresponding LED. So if we press a switch, thereby putting a logic "0" on the

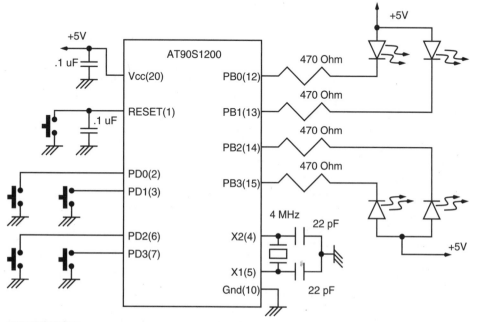

Figure 6.2 Controlling LEDs with switches.

corresponding PORTD pin, we output logic "0" on the matching LED on the PORTB pin. Thus a LED will glow if you press the corresponding switch, and when you release the switch, the LED will stop glowing.

Each port on the AVR processor has three I/O registers associated with it. These registers are called Data Direction register, Output Latch register, and input buffer. These are referred to as DDRx, PORTx, and PINx respectively. So, for portb, these I/O registers are called DDRB, PORTB, and PINB. To output data onto a PORTB pin, you write to the PORTB, and to read data from a PORTB pin, you read the PINB buffer.

The following program is also available on the CD in the code directory as file ledswich.asm.

```
;ledswich.asm
;4 LEDs on PORTB, 4 switches on PORTD
;PORTD0 SWITCH ——·——> PORTB0 LED
;PORTD1 SWITCH ——·——> PORTB1 LED
;PORTD2 SWITCH ——·——> PORTB2 LED
;PORTD3 SWITCH ——·——> PORTB3 LED
;Press one or more switches and corresponding LEDs will lightup
;assembled using Atmel's avrasm assembler.
;the following .inc file should be placed in the same directory as
;this assembly program
.include "1200def.inc"
.cseg
.org 0
        rjmp    RESET               ;Reset Handle
        rjmp    RESET
        rjmp    RESET
RESET:  ldi r16, 0b11111111         ;load register r16 with all 1's
        out DDRB, r16               ;configure PORT B for all outputs
        ldi r16, 0b00000000         ;load register r16 with all 0's
        out DDRD, r16               ;configure PORTD for all inputs
loopit: in r16, PIND                ;read the state of the pin on PORTD
                                    ;into r16 register
        out PORTB, r16              ;and copy it to PORTB
        rjmp loopit
```

The above piece of code shows how to read a switch and light up an LED. However, the switch interfacing is not proper. Typically, a switch, being a mechanical device, doesn't make a clean contact when it is pressed or released.

Figure 6.3 illustrates the signal bounce when a mechanical switch is released. Similar bounce occurs when a switch is pressed. The bounce can last for several milliseconds as illustrated in the figure. Comparatively, the processor executes instructions much faster, up to 1000 times faster or even more. Given such a disparity, if a program were to read a switch and decide to take some action if it is pressed, then even for a single-switch press, it will end up taking the action many, many times. One cure for this problem is to use external damping components such as an RC delay circuit. A better, cost-saving method, which is more elegant, is to provide the damping in software. This software damping scheme is called debouncing the switch.

The way the switch debouncing is performed is as follows: The processor reads the switch input pin, and when it detects a change of logic from "1" to "0" (for the switch configuration as illustrated in Figure 6.2), it knows that the switch has been pressed. It then calls a delay routine, which is of the order of a few milliseconds, say 20 ms (which is the

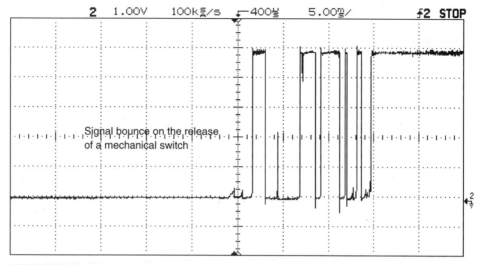

Figure 6.3 **Signal bounce on a mechanical switch when it is released.**

time for which the signal bounce of the switch remains). After this period, the logic on the switch has stabilized to "0". The processor then reads the switch input again to ensure that it is still pressed. The processor then enters a software loop and monitors the switch input pin till the switch is released again. The release of the switch is characterized by the logic at the pin changing from logic "0" to logic "1". After detecting this logic change, the processor again calls a delay routine to timeout the signal bounce on the switch and again checks if the switch has stabilized to a logic value "1". If so, the program concludes that the particular switch was pressed and released and then can take any action as necessary.

Now, before we write a piece of code to read a switch in the way just described, we need to understand how subroutines are written and called in AVR processors.

6.3   Stack Operation in AVR Processors

Subroutine calls are interruptions in the normal sequential flow of the program. To call a subroutine, the address of the subroutine is loaded into the program counter. The processor then starts executing the code resident at this address and onwards. After the subroutine has finished, the program execution must resume from the point where it was suspended in the calling program. To do that, the processor must remember the address of the program memory from where the execution has to resume. This address is stored in a *stack*. A stack is a special storage area that is used to store return addresses. However, the stack is also used for passing parameters to a subroutine, if required, and to return results to the calling program. Typically, the stack is implemented in RAM and is accessed with a special register called Stack Pointer. Stack Pointer is an address register, and it indicates the address of the RAM memory location of the stack.

In the AVR processors, the stack is implemented in two different ways. For those processors that do not have any SRAM, such as the AT90S1200, the processor has a hardware stack.

The hardware stack is three levels deep, meaning that it can store three return addresses. Thus, at any time, only three nested subroutine calls can be made. The hardware stack is only used by the processor to store return addresses. It cannot be used by the program to pass any parameters to the subroutine, as there is no push or pop instruction to access data on the stack. This may seem like a problem, especially for storing the processor state during an interrupt execution. Since the interrupt occurs asynchronously, the state of the SREG register, which has all the flags, can get changed due to instruction execution within the interrupt subroutine. One way out of this is to store the value of the SREG register into another designated register at the beginning of the interrupt subroutine, and while returning from the subroutine, to restore the value from the designated register back into the SREG register as illustrated below.

```
ISR: mov R0, SREG            ;start of interrupt subroutine
                             ;copy SREG value into R0
;Interrupt subroutine code
; ....
; ....
   mov SREG, R0              ;restore R0 value back into SREG
   reti                      ;return from subroutine
```

On the other hand, for those processors that have on-chip SRAM, the processor implements a stack in the SRAM. The stack can be initialized anywhere in this SRAM area. To initialize the stack, the stack pointer is loaded with the address of the SRAM memory, and after this is done, the stack can be accessed by the push and pop instruction. The stack gets used when a subroutine is called or when an interrupt occurs.

The stack grows from a larger memory address into the lower address. Thus, when some data is pushed, data is stored at the current stack pointer address, and then the stack pointer is decremented. Similarly, when the data is popped from the stack, the stack pointer is first incremented and then the data is copied from the stack to the destination register.

Let's now use this information about calling subroutines and improve our lights-and-switches system so that it will now wait for a switch to be pressed, and after a switch is pressed, it will light up the corresponding LED and wait for another switch. If two switches are pressed, then for the one which is pressed earlier, the LED corresponding to that switch will be lit.

The following program is also available on the CD in the code directory as file newswich.asm.

```
;newswich.asm
;4 LEDs on PORTB, 4 switches on PORTD
;PORTD0 SWITCH ——--—> PORTB0 LED
;PORTD1 SWITCH ——--—> PORTB1 LED
;PORTD2 SWITCH ——--—> PORTB2 LED
;PORTD3 SWITCH ——--—> PORTB3 LED
;Press a switch and corresponding to the LED will light up
;press another switch and the first LED will go off and
;the LED corresponding to the new switch will light up
```

```
;assembled using Atmel's avrasm assembler.
;the following .inc file should be placed in the same directory as
;this assembly program
.include "1200def.inc"
.cseg
.org 0
        rjmp    RESET               ;Reset Handle
        rjmp    RESET
        rjmp    RESET
RESET:  ldi r16, 0b11111111         ;load register r16 with all 1's
        out DDRB, r16               ;configure PORT B for all outputs
        ldi r16, 0b00000000         ;load register r16 with all 0's
        out DDRD, r16               ;configure PORTD for all inputs
        ldi r16, 255                ;all LEDs off
        out PORTB, r16
loopit: rcall get_switch            ;call the subroutine to
                                    ;determine which switch is pressed.
                                    ;the subroutine returns the result
                                    ;in register r17
        out PORTB, r17              ;output the value on PORTB
        rjmp loopit                 ;get more
;————————**********————————————-
;GET_SWITCH: Subroutine to determine which switch is pressed.
;switch on        return value in r17
;   PD0          0b11111110
;   PD1          0b11111101
;   PD2          0b11111011
;   PD3          0b11110111
;registers destroyed: r18, r19
;subroutines called: delay20ms
;————————**********————————————-
get_switch:
        in r18, PIND                ;read PIND buffer
        andi r18, $0F               ;
        cpi r18, $0F                ;if no switch is pressed
                                    ;then loop back till pressed
        breq get_switch
        cpi r18, 0b00001110         ;check is SW0 is pressed
        brne not_0                  ;if not check more
its_0:  rjmp next_step
not_0:  cpi r18, 0b00001101         ;check is SW1 is pressed
        brne not_1                  ;if not check more
its_1:  rjmp next_step
not_1:  cpi r18, 0b00001011         ;check is SW2 is pressed
        brne not_2                  ;if not check more
its_2:  rjmp next_step
not_2:   cpi r18, 0b00000111        ;check is SW3 is pressed
        brne get_switch             ;if not some problem, so go back
next_step:
        rcall delay20ms             ;call a debounce delay routine
waitfor_rel:                        ;now wait for the switch to be
        in r19, PIND                ;be released
        andi r19, $0F               ;when the switch is released, all
        cpi r19, $0F                ;PIND0-3 bits will be '1'
        brne waitfor_rel
        rcall delay20ms             ;OK, the switch is released
                                    ;debounce it
        mov r17, r18                ;put the switch code in r17
        ori r17, $F0
        ret                         ;and return
```

```
;——————————**********——————————-
;DELAY20MS: A 20ms delay subroutine
;Crystal Frequency is 4MHz
;registers destroyed: r21, r20
;——————————**********——————————-
delay20ms:
        ldi r21, 31
outer_loop:
        ldi r20, 255
inner_loop:
        nop
        nop
        nop
        nop
        nop
        nop
        nop
        dec r20
        brne inner_loop
        dec r21
        brne outer_loop
        ret
```

6.4 Implementing Combinational Logic

In a previous chapter we mentioned how a controller can be used to implement a simple combinational logic equation as illustrated in Figure 1.3 in Chapter 1.

The figure shows four inputs connected to the PB0, PB1, PB2, and PB3 pins. The output of the circuit is on pin PB4. The following program will implement the logic equation:

```
Output = ((/A * B) + (/B * A)) * (C * /D)
```

The following program (also available on the CD as combi.asm) is only to illustrate how the AVR can be used to implement combinational logic. The program is not optimized. For example, the required output needs an XOR between two inputs, which the AVR can perform. However, I have chosen to implement the XOR using NOT, AND, and OR instructions.

```
;combi.asm
.include "1200def.inc"
.def A=r16
.def Abar=r17
.def B=r18
.def Bbar=r19
.def C=r20
.def Dbar=r21
.def temp=r22
.cseg
.org 0
rjmp RESET                    ;reset handle
RESET:    ldi temp, 0b00001111;
```

```
                out DDRB, temp       ;PB0-3 are inputs
                                     ;PB4-7 are outputs
    loop_here:  in temp, PINB        ;read PORTB pins
                mov A, temp
                mov Abar, temp
                com Abar             ;invert A
                mov B, temp
                mov Bbar, temp
                com Bbar             ;invert B
                mov C, temp
                mov Dbar, temp
                com Dbar             ;invert D
                andi A, 1            ;isolate the bit for A
                andi Abar, 1         ;isolate the bit for Abar
                lsr B                ;get input B to position bit0
                lsr Bbar
                lsr C                ;get input C to position bit0
                lsr C
                lsr Dbar             ;get input D to position bit0
                lsr Dbar
                lsr Dbar
                andi B, 1
                andi Bbar, 1
                andi C, 1
                andi Dbar, 1
                and A, Bbar          ;A = A * Bbar
                and B, Abar          ;B = Abar * B
                and C, Dbar          ;C = C * Dbar
                or A, B              ;A = (A * Bbar) + (Abar * B)
                and A, C             ;A = ((A * Bbar) + (Abar * B))* (C  * Dbar)
                and A, 1
                cpi A, 1
                breq Its1
                cbi PORTB, 4         ;no its 0, so reset PB4
                rjmp loop_here
    Its1:       sbi PORTB, 4
                rjmp loop_here
```

6.5   Connecting the AVR to the PC Serial Port

Now that we have written and tested a couple of programs, it is time to connect the AVR processor to the PC. The simplest port to connect to, on the PC, is the RS-232 serial port. You may want to read the operation of the RS-232 port in detail, presented in a later chapter.

Many of the AVR processors are equipped with a built-in serial port. On the entry-level processors such as the AT90S1200, one can create a software-driven serial port.

This section presents both of these methods. Of course, it is very easy to use the built-in serial port of the AVR processor with only a few instructions. The processor takes care of serializing and shifting out the data on the output pin and assembling the incoming data into a byte. The user needs to set the serial port parameters such as the baud rate (which indicates the bits per second), the number of bits in a transmission, number of stop bits, and parity bit. The processor can generate most of the standard and popular baud rates with a suitable clock frequency.

The serial port of the AVR cannot be connected to the PC serial port rightway. The RS-232 signals are bipolar and in the range of +12 V and −12 V, while the AVR can only handle TTL-level signals (if powered from a +5-V supply). Also, the data as appears on the RS-232 line is inverted. That is to say that when the PC wants to send a logic "0", the voltage on the RS-232 line is +12 V, and when the PC wants to send out logic "1", the line voltage is −12 V. So some sort of RS-232 line driver and receiver that converts the RS-232 signal levels to TTL, and vice versa, is needed. Also, performing the signal inversion is needed.

A very popular RS-232 line driver and receiver that I have extensively used is MAX232 from Maxim, as well as the pin-compatible ADM232 from Analog Devices. The circuit schematic for the RS-232 interface is illustrated in Figure 6.19, appearing later in this chapter.

The following piece of code shows how to set up the built-in serial port (called UART in the AVR datasheets) of the AVR processors. The following program is also available on the CD in the code directory as file uartdrv.asm.

The code is executed on the circuit illustrated in Figure 6.4. I have chosen to use the AT90S8515 for this exercise. The 8515 is connected to the PC serial port through a MAX232 level translator chip. On Windows or DOS, run any terminal emulation program and set the baud rate to 9600, 8 data bits, 1 stop bit, and no parity format. Now type any key; the 8515 will light up the ASCII code of the key on the eight LEDs and also increment the code and transmit it back. So if you press "A", it will send back "B" and so on.

```
;uartdrv.asm
;
.include "8515def.inc"
.def rtemp=r17             ;temporary register
.def rreg=r18              ;register for receiving data
.def treg=r19              ;register for transmitting data
.equ baudrate=$33          ;baud rate of 9600 bps for a clock fre-
                           quency of 8 Mhz
.equ RXC=7                 ;UART receive complete flag( 7th bit of
                           USR register)
.equ UDRE=5                ;UART data register empty flag(5th bit of
                           USR register)
.cseg
.org 0
rjmp RESET                 ;reset handle
rjmp RESET
rjmp RESET
RESET:  ldi r16, low(RAMEND) ;initialize stackpointer
out SPL,r16
ldi r16, HIGH(RAMEND)
out SPH,r16
ldi r16,255                ;initialize port B for output
out DDRB,r16
rcall init_uart            ;initialize 8515 for transmit and receive
up: rcall rxcomp           ;receive a byte of data
     mov treg, rreg
     com rreg
out portb,rreg             ;output the data on port B
inc treg                   ;increment the received byte
rcall txcomp               ;transmit the byte
  rjmp up
;*********************************************************
;INIT_UART: Initialize the UART for 9600 bits per second
;        8 data bits, 1 stop bit, no parity
;*********************************************************
```

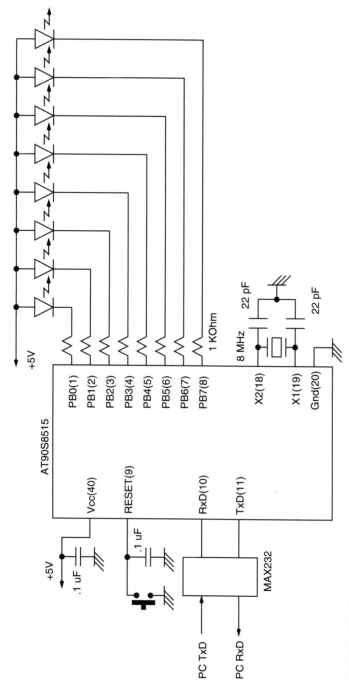

Figure 6.4 Connecting AT90S8515 to a PC serial port. Other components that go with MAX232 are not illustrated.

```
init_uart:
        ldi rtemp, baudrate      ;set baud rate
        out UBRR,rtemp
        ldi rtemp, $18           ;initialize UART control register
        out UCR, rtemp
        ret
;************************************************************
;RXCOMP: Receive a byte from the serial port
;polls the RXC flag in the UART Status Reg (USR)
;if '1', then data is read from the UART Data Register (UDR)
;************************************************************
rxcomp: sbis USR,RXC             ;poll to check if char received
        rjmp rxcomp
        in rreg,UDR              ;put received data in rreg
        ret
;************************************************************
;TXCOMP: Transmit a byte from the serial port
;polls to see if the UDRE flag is '1'. If '1' then
;a byte is written to the UDR to be transmitted.
;************************************************************
txcomp: sbis USR,UDRE            ;poll to check end of transmission
        rjmp txcomp
        out UDR, treg
        ret
```

For those processors that do not have a built-in UART, we describe a software-driven serial port. A software-driven serial port can only be half duplex, meaning that either the serial data can be received or it can be transmitted. A hardware UART, on the other hand, can be full duplex, as the data transmission and reception are being handled by hardware registers that do not need any program intervention in the actual bit-shifting process.

Figure 6.5 illustrates the timing involved in a RS-232 transmission. To receive a serial bit stream, the program must monitor the signal (the TTL signal as illustrated in the figure). The idle state of the serial TTL signal is "1". As soon as a low-going transition is detected, it denotes the beginning of the Start bit and the start of a transmission. The program just monitors the signal again at T/2 time later, which is denoted as the 0th sample. T denotes the bit time. For a 2400-bps speed, the bit time T = 1 / 2400, which is about 416 us. Thus after ensuring that the signal is still "0", the program then just samples the TTL signal at each T time interval after the 0th sample at sample points denoted by 1, 2, 3, etc. The program just records the logic at these sample intervals and shifts the recorded logic values in a register. At the end of eight sample points, the data byte is ready.

Serial data transmission is easy compared to receiving it. The program just generates a start bit for T time units and then shifts out the data to be transmitted, each bit lasting T time units. To get the timing intervals, the AVR processor can use the Timer0 timer, which is available in all the AVR processors.

The software-driven data transmission and reception routine that is included on the CD in fact interfaces directly to the RS-232 port without using any MAX232 type of line converters. The level conversion from RS-232 level to TTL is performed with a few resistors and diodes. The signal inversion is performed in software. For transmission, the TTL data can be directly put on a RS-232 line, and the PC will receive it correctly (the data must be inverted in logic, though).

A working example of this approach is illustrated in a later chapter, and the circuit diagram is illustrated in Figure 17.7. Serial driver (bit-banging method) test code for Figure 17.7 is available in the code directory in file ser_drv.asm.

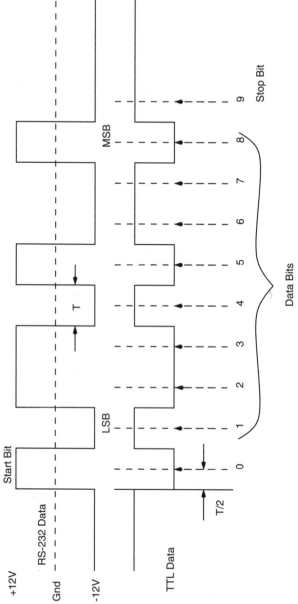

Figure 6.5 Timing the RS-232 signal. The first bit is the Start bit and the last bit is the Stop bit.

6.6   Expanding I/O

The AVR processors are available many different pinouts, with different I/O resources, depending upon the number of pins in the particular processor. In some cases, you may feel the need for additional I/O pins than are available. There are many ways to expand the number of I/O pins with the help of shift registers or port expanders with a SPI or I2C interface. This section discusses means of expanding I/O.

The primary requirement is that the I/O expansion scheme should have some serial format so as to take up minimum I/O pins on the processors. Serial Shift registers are great for such applications. Usually, these shift registers are of the serial-in and parallel-out or parallel-in and serial-out format, which suits our requirement. There are many bidirectional I/O expansion ICs with a 2-wire I2C interface available that are, of course, the best in terms of minimum pin usage.

6.6.1   I/O EXPANSION USING SHIFT REGISTER

Figure 6.6 illustrates the scheme for a an 8-bit digital input port using an 8-bit parallel-in, serial-out shift register. This expansion scheme requires 3 I/O pins, and for the cost of 3 I/O pins, you get 8 input-only pins. The 74165 has 5 control lines: serial-in to cascade multiple shift registers, Qout, which is the shift register output, Clock Inhibit to disable clocking of the shift register, Shift/Load*, that is used to capture the input data and shift it out through the Qout pin, and the Clock input pin.

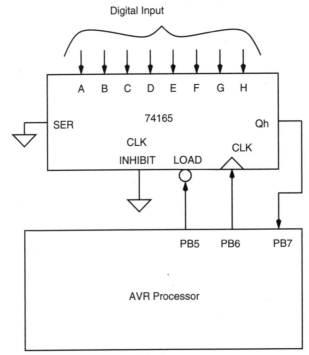

Figure 6.6 Eight-bit digital input port using a parallel-in serial-out shift register.

For an 8-bit input port, we need just one 74165, and so the serial-in pin is connected to ground. The clock inhibit pin is also grounded so the clock input is always enabled. The Qout pin is connected to the PORTB7 pin for reading in the shift register data, the Clock signal pin is connected to the PORTB6 pin, and the Shift/Load* pin is connected to the PORTB5 pin.

To read a byte of input data from this expansion port, the Shift/Load* pin is reset to "0" momentarily and then set to "1". This captures the input data in an internal register in the shift register. After this, the Clock signal is pulsed and for each pulse, the PORTB7 pin is read and a bit is shifted out in an internal register. After eight such clock pulses and shifts, the entire byte from the 74165 shift register is read into the AVR processor.

Similarly, Figure 6.7 illustrates an 8-bit output only port. The circuit operates similar to the input port expansion scheme, except that the PORTB7 pin is used to output data to the output shift register CD4094. Eight bits of data are shifted into CD4094, and after eight shifts, the strobe signal for the output stage latch of the CD4094 is set to "1" to transfer the shift register data to the output pins. When the data is being shifted into the shift register, the strobe signal is held at logic "0".

6.6.2   IIC EXPANDERS

In addition to the shift register method of expanding the I/O capacity of an AVR processor, there exists another method to expand I/O capacity. The idea is to use IIC bus-based I/O expander ICs. Manufacturers have perceived the need for increasing the I/O and have designed chips for the purpose. Philips, who is the developer of the IIC bus has designed

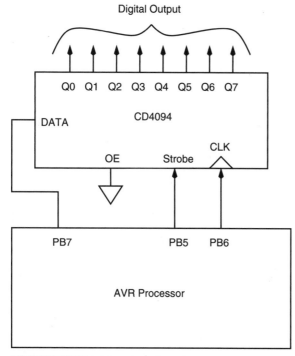

Figure 6.7 Eight-bit digital output port using a serial-in parallel-out shift register.

many IIC I/O expanders. Figure 6.8 illustrates the block diagram of just such an I/O expander. It offers one 8-bit bidirectional port. Up to eight such ICs can be hooked on the same IIC bus to achieve more I/O capability.

Figure 6.9 illustrates how the PCF8574 I/O expander IC can be connected to the AVR processor. The INT* output is connected to the INT0 input of the AVR so that by sending an interrupt signal on this line, the remote I/O can inform the microcontroller if there is incoming data on its ports without having to communicate via the I2C-bus. This means that the PCF8574 can remain a simple slave device.

6.7 Interfacing Analog-to-Digital Converters

An analog-to-digital converter (ADC) is a device that converts analog voltage to a digital number. An ADC is used to digitize analog signals. A signal varying with time is sampled at discrete time intervals, and a number representing the amplitude of the signal at the instant is recorded. This is illustrated in Figure 6.10. The code output is on the Y axis and the time is on the X axis. The code output has eight levels, and these can be encoded with three bits. So the encoded binary number ranges from 000 to 111.

There are many types of ADC techniques, and we will not go into those details. I will mention the type of ADC when we consider a particular chip. For now, let's see how the AVR processor can be used to encode an external analog signal.

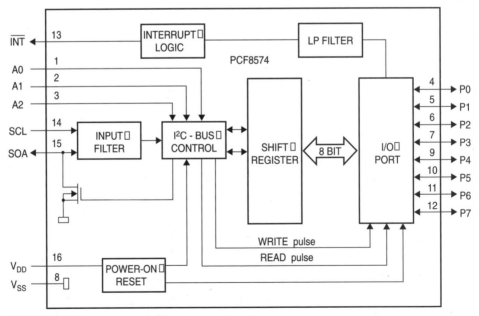

Figure 6.8 Eight-bit bidirectional digital I/O port expander.

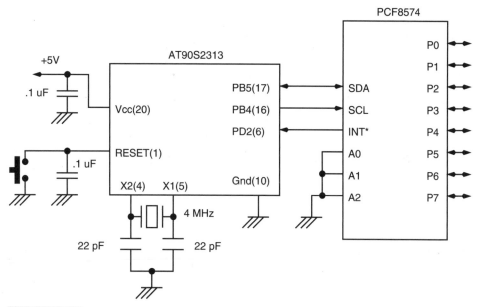

Figure 6.9 AVR interface to PCF8574.

6.7.1 AD CONVERSION USING THE ON-CHIP COMPARATOR

The simplest ADC can be built with the AVR processor by using the on-chip analog comparator together with either a timer or even a counter. (See Figure 6.11.) Figure 6.12 illustrates a rather crude ADC using the on-chip analog comparator. The comparator compares the voltages on the +ve input Ain0 and the −ve input Ain1 and if the Ain0 voltage is greater (i.e., more positive) than the Ain1 voltage, the output of the comparator ACO is set to "1". The ACO output is directly readable as a bit in the ACSR register.

The simple ADC in Figure 6.12 works as follows. To begin with, the PB0 pin is set to logic "0". This discharges any charge on the capacitor. Then the PB0 is pin programmed as an input with no pull-up resistors, and either a software counter or Timer0 (or Timer1) is triggered to start counting. The capacitor starts charging to +5 V through the resistor R1. When the voltage on the capacitor becomes more than input voltage on the Ain1 pin, the comparator output switches "1". When this is detected by the program, which is polling in a loop to detect the change of state of the comparator to "1", the software counter (or the Timer0) is stopped and the accumulated count is proportional to the input voltage on Ain-pin. The larger the voltage, on Ain-, the capacitor will have to charge to a voltage higher than that, and that will take more time, which means the internal counter will be clocked for more time, accumulating a larger count. There is only one glitch (and a rather undesirable one) in this approach, and that is the voltage on the capacitor does not increase linearly but exponentially. So the accumulated count is not linearly proportional to the input voltage. However, by restricting the input to a small range of say 0 to 2.5 V, a fairly linear region of the RC charging curve would be used. A normalized plot of the difference between the count generated by the RC charging method and the true count is provided by the plot illustrated in Figure 6.13. The input range of voltage is restricted between 0 and 2.5 V.

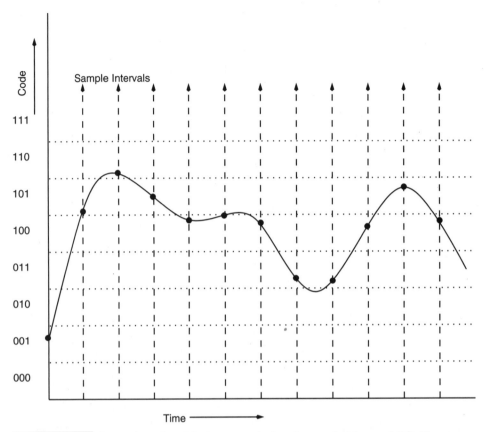

Figure 6.10 An analog signal being sampled and encoded by an ADC. The number output of the ADC is on the Y axis and the time is on the X axis.

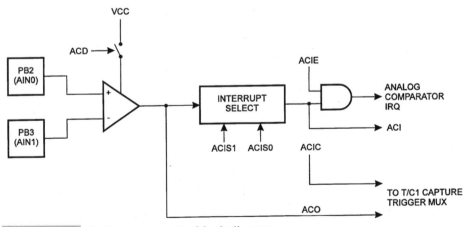

Figure 6.11 Analog comparator block diagram.

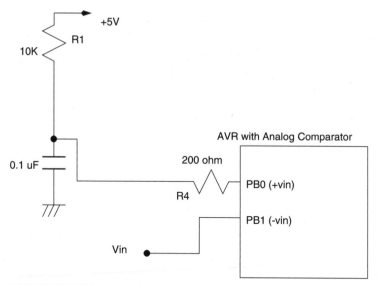

Figure 6.12 Block diagram for a crude analog-to-digital converter using the on-chip comparator on an AVR processor.

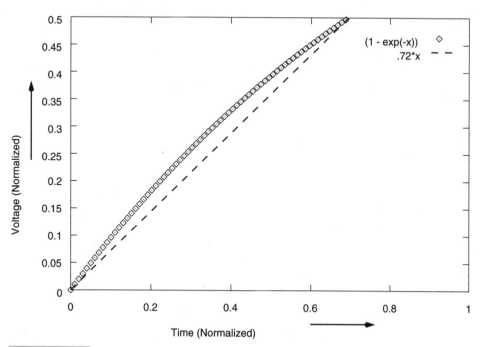

Figure 6.13 A linear and an exponential plot for a small input range. This plot gives an idea of the amount of nonlinearity between the count accumulated using the simple RC charging scheme and the ideal count.

For the values illustrated, the RC time constant is 1 ms, and so the capacitor will charge to 2.5 V in about 720 μs. Thus the worst-case conversion time by the scheme is 720 μs.

The nonlinearity in the RC charging can be removed with the help of a scheme as illustrated in Figure 6.14, where the resistor R1 is replaced with a transistor current source. A constant current source will charge the capacitor linearly, and so the accumulated count will be linearly proportional to the input-applied voltage on the Ain-pin.

With the values illustrated in Figure 6.14, the current sourced by the PNP transistor is about 80 μA. The charging of a capacitor with a constant current source is expressed with the equation:

$$dv/dt = I/C$$

Therefore, the time to charge a capacitor from 0 to 5 V is

$$T = (C * 5) / I$$

Plugging in the values, T = 6:25 ms. Thus the worst-case conversion time with this scheme is 6.25 ms. This can be modified by changing the current provided by the current source.

The conversion time is long enough to extract a 10-bit or even 12-bit resolution—i.e., a 10- or 12-bit counter (either software or with the help of Timer0 or Timer1) can be filled up easily during the conversion time.

The resistor R4 is used to limit the capacitor discharge current to within safe values. Without this resistor, the PB0 input would get damaged from the capacitor discharging a large transient pulse into the PB0 pin.

One of the possible applications of the improved ADC is as a temperature recorder. Figure 6.15 illustrates the circuit. LM335 is a temperature sensor that provides a voltage output proportional to the ambient temperature. The voltage generated by the sensor is 10 mV/K. This temperature operates from −40°C to +100°C temperature. So at room tem-

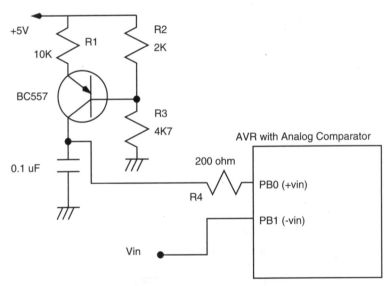

Figure 6.14 Block diagram for an improved analog-to-digital converter using the on-chip comparator on an AVR processor.

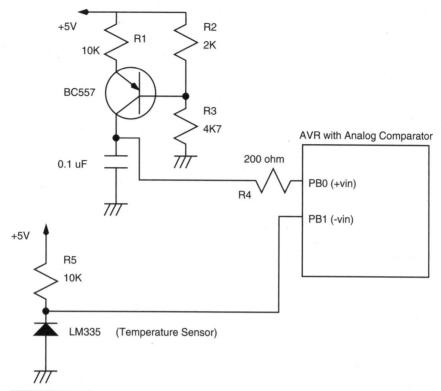

Figure 6.15 Block diagram for a temperature sensor interface to the comparator-based ADC.

perature (20°C, which is 393 K), the voltage would be 3.93 V. A simple AT90S1200 can be used for this application and the data can be either stored in the internal EEPROM or transmitted to a PC using the software-driven serial link as described in a previous section.

6.7.2 MAX186

MAX186 is a 12-bit, 8-channel serial ADC system with built-in voltage reference, internal sample and hold amplifier, and multiplexer. The ADC offers various modes of operation such as single-ended conversion, differential conversion, sleep mode, etc. The maximum current conversion is 2 mA and 100 μA during low power modes (sleep mode).

MAX186 is a complete ADC system combining an 8-channel analog multiplexer, sample and hold amplifier, serial data transfer interface and voltage reference, and a 12-bit resolution successive approximation converter. All of these features are packed into a 20-pin DIP package (other packaging styles are also offered). The IC consumes extremely low power and offers power-down modes and high conversion rates. The power-down modes can be invoked in software as well as hardware. The IC can operate from a single +5-V as well as ±5-V power supply.

The analog inputs to the ADC can be configured via software to accept either unipolar or bipolar voltages. The inputs can also be configured to operate as single-ended inputs or differential inputs. The ADC has an internal voltage reference source of 4.096 V, but the user can choose not to use this reference and supply an external voltage between +2.50 V and +5.0 V. This gives the user the advantage of adjusting the span of the ADC according to the need—e.g., if the input analog voltage is expected to be in the range of 0 to +3.0 V, then choosing a reference voltage of 3.0 V will provide the user with the entire ADC input range with a better resolution.

This ADC is an extremely fast device. It can convert at up to 133000 samples per second at the fastest serial clock frequency. This ADC is best suited for devices that can generate fast serial-controlled clocks—e.g., DSPs and microcontrollers such as the AVR. Figure 6.16 illustrates the block diagram of the ADC and the various associated signals. The description of the ADC signals is listed in Table 6.1.

6.7.3 MAX186 DATA CONVERSION AND READOUT

While the many details of this very fine ADC can be had from the IC manufacturer data sheets, our intention here is to see how we can connect this device to the AVR processor to begin with and how a conversion can be initiated and the result read out into the AVR processor.

To initiate a conversion, the ADC must be supplied with a control byte. The control byte is input into the ADC through the Din signal input. To clock the control byte, either an internally or externally generated clock signal (on SCLK pin) could be used. To keep the hardware small and simple, it is necessary to use the external clock mode. The format of the control byte is illustrated in Figure 6.17.

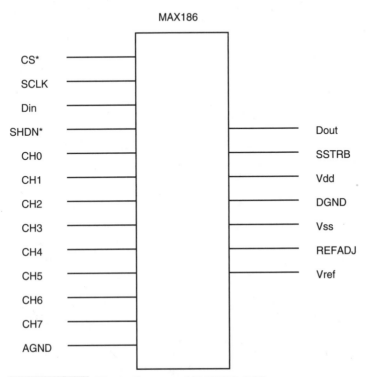

Figure 6.16 Block diagram of MAX186 ADC.

TABLE 6-1 ADC MAX186 SIGNALS AND THEIR FUNCTIONS

SIGNAL NAME	FUNCTION
CS*	Active low-chip select input
SCLK	Serial clock input. Clocks data in and out of the ADC. In the external clock mode, the duty cycle must be 45% to 55%.
Din	Serial data input. Data is clocked at the rising edge of SCLK.
SHDN*	Three-level shutdown input. A low input puts the ADC in low-power mode and conversions are stopped. A high input puts the reference buffer amplifier in internal compensation mode. A floating input puts it in external compensation mode.
CH0-CH7	Analog inputs.
AGND	Analog ground and input for single-ended conversions.
Dout	Serial data output. Data is clocked out at the falling edge of SCLK.
SSTRB	Serial strobe output. In external clock mode, it pulses high for one clock period before the MSB decision.
DGND	Digital ground.
Vdd	Positive supply voltage. +5 volts ±5%.
Vss	Negative supply voltage. −5 volts ±5% or AGND.
REFADJ	Input to the reference buffer amplifier.
Vref	Reference voltage for AD conversion. Also output of the reference buffer amplifier (+4.096 volts). Also, input for an external precision reference voltage source.

To clock the control byte into the ADC, the CS* pin is pulled low and a rising edge on SCLK clocks a bit into Din. The control byte format requires that the first bit to be shifted in should be "1". This defines the beginning of the control byte. Until this start bit is clocked in, any number of "0" can be clocked in by the SCLK signal without any effect.

The control byte must be 1XXXXXX11 (binary). Xs denote the bits required for channel and conversion mode selection. The two least-significant bits are set to '1' and "1" to select the external clock mode option.

Figure 6.17 illustrates the control byte format. The control byte value for starting a conversion on channel 0 of the ADC, in unipolar, single-ended conversion mode using external clock, is 10001111 (binary) or 8F hex.

Let's now consider the timing diagram in Figure 6.18, which illustrates the conversion and readout process on ADC channel 0.

The timing diagram illustrates five traces, namely CS*, the chip select signal; SCLK, the serial clock required for programming the ADC and the subsequent readout; the Din, which carries the programming information (the control byte); the SSTRB, which the ADC generates to indicate the beginning of the readout process; and Dout, the actual data output from the ADC, which is the conversion result.

The data on signal Din is clocked into the ADC at the rising edge of the SCLK signal.

MSB							LSB
START	SEL2	SEL1	SEL0	UNI/BP*	SGL/DF*	PD1	PD0

START: The first logic '1' bit after CS* goes low defines the start of the Control byte

SEL2, SEL1, SEL0: These 3 bits select which of the 8 channels will be used for conversion

UNI/BP*: 1=Unipolar; input can range between 0 to +Vref;

0=Bipolar; input can range between +Vref/2 to -Vref/2

SGL/DF*: 1=single ended; 0=Differential

PD1, PD0: Defines clock & power down modes.

0   0 : Full power down mode

0   1 : Fast power down mode

1   0 : Internal clock mode

1   1 : External clock mode

Figure 6.17 MAX186 control byte format.

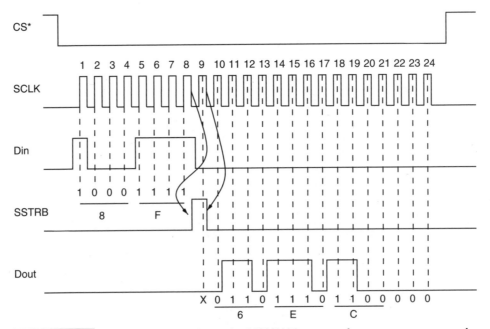

Figure 6.18 Timing diagram of a typical MAX186 conversion process as record-ed on a logic analyzer.

The first bit that is clocked in is D7. To begin the conversion, D7 needs to be set to "1", as can also be seen from the value of the control byte that we calculated. So Din is set to "1" and the first SCLK rising edge is applied to the ADC. The SCLK is then taken low.

Thereafter, the Din is set to each of the subsequent bits of the control byte before applying the SCLK. At the end of 8 SCLK pulses, the Din bit is not required and is set to "0". At the falling edge of the 8th SCLK pulse, the ADC sets the SSTRB bit to "1". At the falling edge of the 9th SCLK bit, SSTRB is taken to "0".

At the falling edge of the 9th SCLK signal, the ADC outputs data on the Dout signal, one bit for each of the next 15 falling edges of the SCLK signal. The data on the 9th pulse is "0" and the actual conversion result is effective after the 10th falling edge to the 21st rising edge. Thereafter, for the next 3 edges, the ADC outputs "0"s.

For a controller circuit such as the AVR, with minimal parts, to initiate conversion and readout the result would need three output bits and one input bit. The output bits would be needed to generate the Din and SCLK signal and the input bit to read the Dout signal from the ADC.

Figure 6.19 illustrates the circuit with a MAX186 ADC, an AT90S2313 processor, and a MAX232 RS-232 level translator. The circuit is connected to the PC serial port. The AVR processor waits for a command from the PC and then initiates conversion on the MAX186 and sends out the data back to the PC serial port.

The program is available on the CD. The code for this project is available in the code directory in the file MX186_ex.asm.

6.7.4 MAX110/MAX111

MAX111/MAX110 is a serial 14-bit, dual-channel ADC from Maxim. MAX111-/MAX110 ADC uses an internal autocalibration technique to achieve 14-bit resolution without any external component. The ADC offers two channels of ADC conversion and operates with 650µA current, thus making it ideal for portable, battery-operated data acquisition operations.

MAX111 operates from a single +5-V power supply and converts differential signals in the range of ±1.5 V or differential signals in the range of 0 to 1.5 V.

MAX111 can operate from an external as well as internal oversampling clock that is used for the ADC conversion. To start a conversion, digital data is shifted into the MAX111 serial register after pulling the CS low. CS can only be pulled low when BUSY is inactive. MAX111 has a fully static serial I/O shift register which can be read at any serial clock (SCLK) rates from DC to 2 MHz. Input data to the ADC is clocked in at the rising edge of the SCLK and the output data from the ADC (conversion result) is clocked out at SCLK falling edge and should be read on SCLK rising edge.

The data clocked into the ADC determines the ADC operation, which could be to initiate a new conversion, calibrate the ADC, perform offset null, change ADC channel, change oversampling clock divider ratio, etc.

The format of this control word is as follows:

```
bit #  15      14      13      12      11      10      9      8
       No-op   NU      NU      CONV4   CONV3   CONV2   CONV1  DV4
bit #  7       6       5       4       3       2       1      0
       DV2     NU      NU      CHS     CAL     NUL     PDX    PD
```

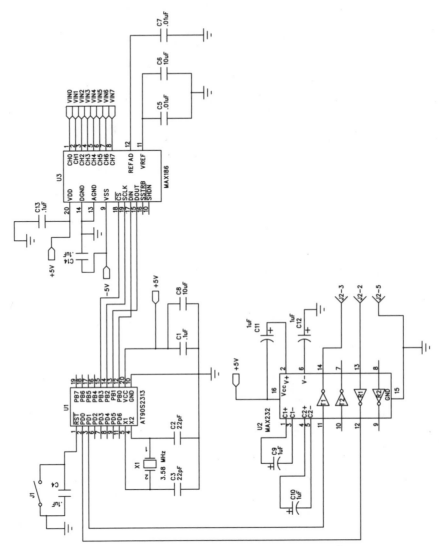

Figure 6.19 Circuit schematic for an AT90S2313 processor interface to the MAX186 ADC.

BIT NAME	FUNCTION
No-OP	If this bit is 1, the remaining 15 bits are transferred to the control register and a new conversion begins when CS* returns high.
NU	Not Used, should be set low.
CONV1-4	Conversion time control bits.
DV4-2	Oversampling clock ration control bits.
CHS	Input channel select, logic 1 selects channel 2, low selects channel 1.
CAL	Gain Calibration bit. A high bit selects gain calibration mode.
NUL	Internal Offset Null bit. Logic high selects this mode.
PDX	Oscillator power down bit, selected with logic high.
PD	Analog power down bit selected with logic high.

Figure 6.20 illustrates an AVR processor interfaced to the MAX111 ADC.

The AVR controller monitors the status of BUSY* signal, which indicates if the ADC is busy with a conversion. A "0" on this pin indicates that the ADC is still converting. The program reads the status of BUSY* on the PORTD2 pin. When the program finds BUSY* at logic "1", it pulls the CS* signal of the ADC low to start a new conversion process.

It then generates 16 clock pulses on the PORTD5 pin connected to the SCLK signal pin of the ADC. Synchronized to these pulses, the program generates a serial bit stream on pin PORTD4 connected to the Din pin of the ADC. This bit stream contains the control

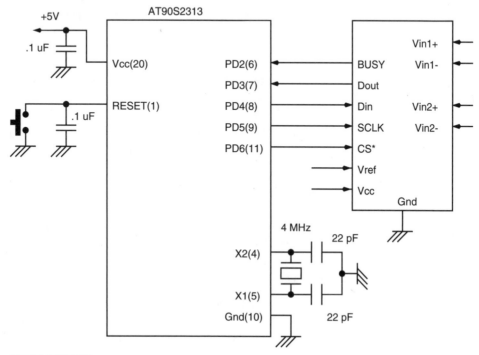

Figure 6.20 MAX111 interface to the AVR processor.

word with the format described previously. Output data from the ADC is clocked out on Dout pin on the falling edges of the SCLK pulses. The program reads this data on the PORTD3 pin. The CS* signal connected to the PORTD6 pin is pulled up after the 16 clock pulses are generated.

The ADC pulls its BUSY* signal low while the conversion is in progress. The conversion time depends upon the XCLK frequency and the format of the control word. In this circuit, the internal RC oscillator is used for the conversion clock. The converted data is clocked out in the next round of the clocking sequence by the ADC.

Figure 6.21 illustrates the timing diagram of a typical conversion and readout sequence recorded on a logic analyzer. A suitable data conversion and readout driver code is included in a later project chapter. The driver program is in "C".

6.8   Interfacing Digital-to-Analog Converters

Digital-to-Analog Converters (DACs) are devices that function exactly opposite to the ADCs. DACs convert digital data to analog voltage (or current). Functionally, the DAC has n digital input lines and 1 output line that provides analog voltage or current. The analog output is proportional to the weighted sum of the digital inputs.

6.8.1   USING PWM FOR A DAC

Pulse Width Modulation (PWM) technique can be used easily to create a DAC, especially since many of the members of the AVR processor family are equipped with on-chip PWM.

In PWM, a digital signal of fixed frequency is generated. The pulse width of the signal is changed according to the requirement. Ideally, it should be possible to vary the width to any arbitrary value. However, with a counter-based PWM, the change can be only as much as the resolution of the counter. A PWM implemented using an 8-bit counter can only change the pulse width by 0.4% approximately (1 bit in 255). By employing a low-pass filter at the output of a PWM wave, the average value of the signal is extracted. The average value of a digital signal is equal to the duty cycle of the waveform.

Figure 6.22 illustrates a 2-bit PWM signal. Figure 6.23 illustrates a DAC using the built-in PWM generator on output PORTB3 pin and an external RC filter. The RC filter has a bandwidth of about 16 Hz for the values illustrated. The Timer1 can be clocked at the system clock frequency of 4 MHz as illustrated, and for an 8-bit PWM, the PWM frequency will be about 7800 Hz. The RC filter will cleanly filter out the high-frequency components, and a clean DC value will be produced.

6.8.2   R-2R LADDER DAC

Using only two different values of resistors, it is possible to build a simple R-2R ladder DAC of reasonable linearity. Figure 6.24 illustrates a R-2R ladder DAC connected to the PORTB of the AT90S1200. To use the DAC, the following code can be used.

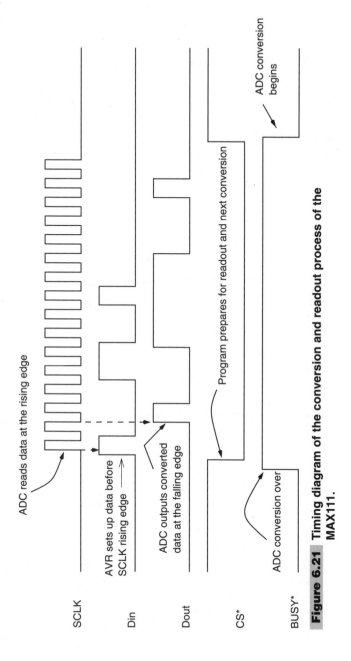

Figure 6.21 Timing diagram of the conversion and readout process of the MAX111.

SCLK

ADC reads data at the rising edge

Din

AVR sets up data before SCLK rising edge —→

Dout

ADC outputs converted data at the falling edge

CS*

Program prepares for readout and next conversion

BUSY*

ADC conversion over

ADC conversion begins

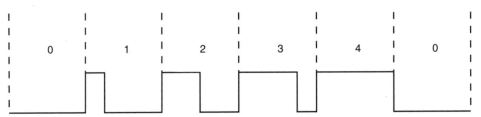

Figure 6.22 A continuously varying PWM signal. The average value of the signal changes by 25% in each period.

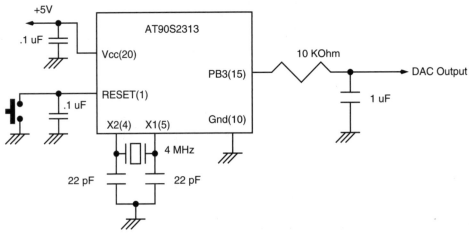

Figure 6.23 PWM DAC using an AT90S2313 and an output RC filter.

```
.include "1200def.inc"
.def DACVALUE=r17              ;Register with the DAC value
.def temp    =r18
init_portb:  ldi temp, $FF     ;Initialize PORTB as output
             out DDRB, temp
load_dac:    out PORTB, DACVALUE ;output value to the DAC
```

6.8.3 MAX521 DAC

MAX521 is a voltage output DAC and has a simple 2-wire digital interface. These 2 wires can be connected to more MAX521s (total up to 4). The IC operates from a single +5-V supply. Even with a +5-V supply, the outputs of the DACs can swing from 0 to +5 V. The IC has 5 reference voltage inputs that have a range that can be set to anywhere between 0 to +5 V.

Figure 6.25 illustrates the block diagram of the IC. Table 6.2 lists the signals of the MAX521 DAC IC. The MAX521 has five reference inputs. The first four DACs each have independent reference inputs, and the last four share a common reference voltage input.

The digital interface allows the IC to communicate to the host at a maximum of 400 Kbps. The input of the DACs have a dual data buffer. One of the buffer outputs drives the DACs while the other can be loaded with a new input. All the DACs can be set to a new value independently or simultaneously. The IC can also be programmed to a low-power mode, during which time the supply current is reduced to 4 μA only. The power-on reset circuit inside the IC sets all of the DAC outputs to 0 V when power is initially applied.

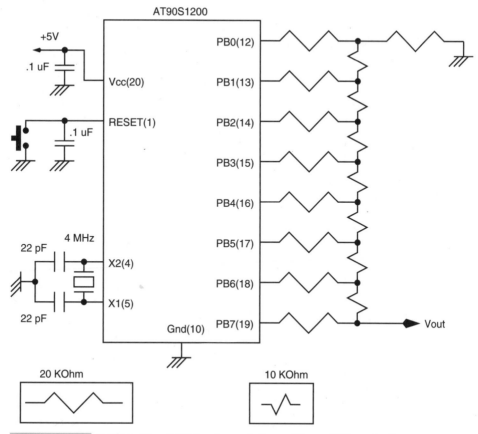

Figure 6.24 R-2R ladder DAC implementation with an AVR controller.

The output of an 8-bit DAC is

$$Vout = Vref \ (input/256)$$

where input is an 8-bit number and Vref is the reference voltage for the channel.

6.8.4 DATA TRANSFER TO A MAX521

The MAX521 uses a simple two-wire interface. Up to four MAX521s can be connected to one set of these two-wire interfaces. This means that a host system with two output lines can be used to program up to 32 DACs!

To send commands and data to MAX521, the host sends logic sequences on the SDA and SCL lines. Otherwise, these lines are held to "1". The two-wire interface of MAX521 is compatible with the I2C interface. To maintain compatibility with I2C, external pull-up resistors on the SDA and SCL lines would be required. Otherwise, these resistors are not required.

MAX521 is a receive-only device, so it cannot transmit any data. The host only needs two output signal lines for SDA and SCL signals. The SCL clock frequency is limited to 400 kHz. The host starts communication by first sending the address of the device followed by the rest of the information, which could be a command byte or a command byte

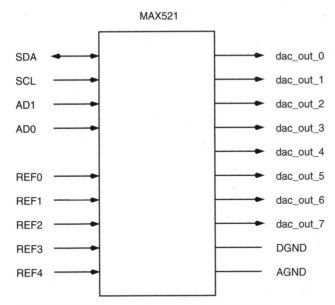

Figure 6.25 Block diagram of MAX521 DAC.

and data byte pair. Each such transmission begins with a START condition as illustrated in the timing diagram in Figure 6.26, followed by the device address (called the slave address) and command-byte, data-byte pairs or command byte alone. The end of transmission is signaled by the STOP condition on the SDA and SCL lines.

The SDA signal is allowed to change only when the SCL signal is low, except during the START and STOP conditions. For the START condition, the SDA signal makes a high to low transition while the SCL signal is high. Data to the MAX521 is transmitted in 8-bit packets (which could be the address byte, the command byte, or the data byte) and it needs nine clock pulses on the SCL signal line. During the ninth SCL pulse, the SDA line is held low, as illustrated in the timing diagram. The STOP condition is signaled by a low to high transition on the SDA signal line when the SCL signal is held high.

The address and command bytes transfer important information to MAX521. The address byte is needed to select one out of a maximum of four devices that could be connected to the SDA-SCL signal lines. After the host starts the communication with the START condition, all the slave devices on the bus (here the bus is referred to the SDA and SCL signal lines) start listening. The first information byte is the address byte. The slave devices compare the address bits AD0 and AD1 with the AD0 and AD1 pins condition on the IC. In case a match occurs, the subsequent transmission is for that slave device.

The next transmission is either a command byte or a command-byte, data-byte pair. In either case, the data byte, if at all, follows the command byte, as illustrated in Figure 6.27. Table 6.3 lists the bit sequence of the command byte and the function of each bit.

All the possible combinations of address byte, command byte, and data byte to a MAX521 are:

TABLE 6-2   SIGNAL DESCRIPTION OF THE MAX521 DAC

SIGNAL NAME	FUNCTION
OUT0	DAC0 voltage output
OUT1	DAC1 voltage output
OUT2	DAC2 voltage output
OUT3	DAC3 voltage output
OUT4	DAC4 voltage output
OUT5	DAV5 voltage output
OUT6	DAC6 voltage output
OUT7	DAC7 voltage output
REF0	Reference voltage input for DAC0
REF1	Reference voltage input for DAC1
REF2	Reference voltage input for DAC2
REF3	Reference voltage input for DAC3
REF4	Reference voltage input for DACs 4, 5, 6, and 7
SCL	Serial Clock input
SDA	Serial Data input
AD0	Address input 0. Sets IC's slave address
AD1	Address input 1. Sets IC's slave address
Vdd	Power supply, +5 volts
DGNC	Digital ground
AGND	Analog ground

1. START condition, slave address byte, command byte/output data byte pair, and a STOP condition, or
2. START condition, slave address byte, command byte, STOP condition, or
3. START condition, slave address byte, multiple command byte/output data byte pairs, STOP condition.

Figure 6.28 illustrates how to connect up to 4 MAX521s on a single bus from the host. The four devices are distinguished by the different addresses set on the AD0 and AD1 lines. Each of the MAX521 compares these bits with the address bits in the address byte transmission from the host.

Figure 6.29 illustrates how an AT90S2313 AVR processor can be connected to a MAX521 DAC to provide up to 8 channels of 8-bit DAC. The I2C protocol that the MAX521 DAC understands can be created on any I/O line of the processor, and the processor can communicate to the MAX521 chip under software control.

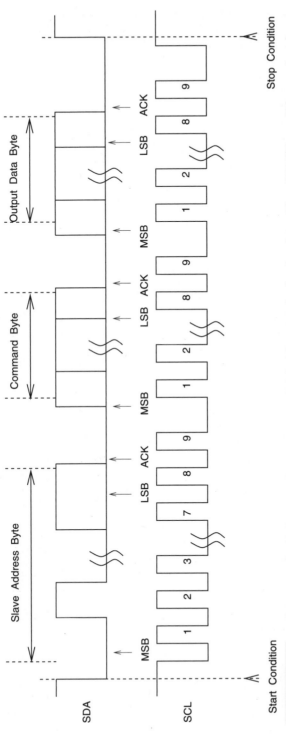

Figure 6.26 Communication format for MAX521 serial DAC. All transmission begins with a START condition and ends with a STOP condition.

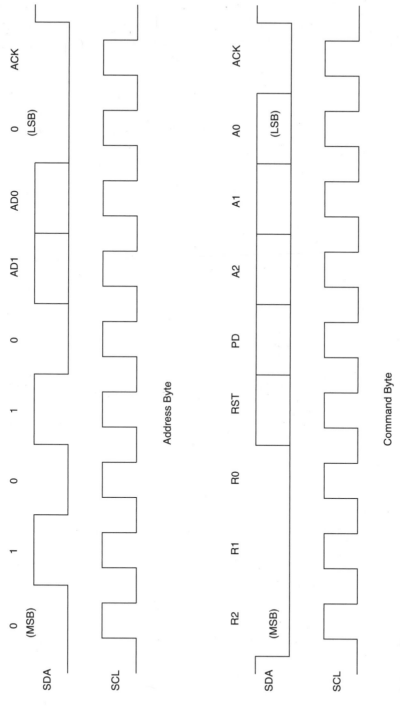

Figure 6.27 Structure of the Address and Command bytes.

131

TABLE 6-3	BITS OF THE COMMAND BYTE FOR MAX521
BIT NAME	**FUNCTION**
R2, R1, R0	Reserved bits. Set to "0."
RST	RESET bit. A "1" on this bit resets all DAC registers.
PD	Power Down bit. A "1" on this bit put MAX521 in a power down mode. A "0" returns the MAX521 to normal state.
A2, A1, A0	Address bits. Defines address of the DAC to which the subsequent data byte will be addressed.
ACK	Acknowledgment bit. Set to "0."

6.9 Interfacing LED Displays

Displays are an important component in an embedded system. They are one of the most popular ways to communicate with the system user. There are many types of display devices that can be used and interfaced with the AVR processor.

6.9.1 SEVEN-SEGMENT DISPLAYS

The simplest display device is, of course, a LED and we have already seen how it can be connected and used with the AVR. But it can provide limited information to the user. A LED seven-segment display, on the other hand, can be used to provide numeric information. It requires eight signal lines if possible and at least seven at the minimum. The display has seven LEDs labeled "a" through "g" and then there is a decimal point. Figure 6.30 illustrates a scheme to connect two LED seven-segment displays to the AVR processor.

This puts an immense resource load on the processor. Interfacing a couple of seven-segment displays takes up all the I/O pins. The situation is remedied by using a multiplex scheme. Here, at the cost of increased software complexity, some of the I/O pins can be saved while at the same time more displays can be added. This scheme has the advantage that our eye cannot follow any light change faster than about 20 Hz. So if an LED display is put on and off at a rate greater than 20 Hz, due to the persistence property of the eye, it will not feel any difference, provided that the average intensity of the LED is maintained. Thus, many LED displays can share the same I/O lines, with only one of them being lit at any time.

Figure 6.31 illustrates how four LED displays can connect to the AVR controller using the multiplex scheme. The power to each display (common anode type display) is controlled by an output signal line of the AVR through a PNP transistor switch. A "0" at the base turns it ON and provides voltage to the display. At any given time, only one of the transistors is turned ON. Once a +ve voltage is applied to a display, the cathodes being connected to port lines, the pattern of "1"s and "0"s on the port will determine which LED segment glows. A "0" on the port will sink the current from the segment and the segment will glow.

The seven-segment LED displays cannot be used to display alphabets (well, only a limited number of alphabets can be displayed). To display aphanumeric information, there are alphanumeric displays available which have sixteen segments as illustrated in Figure 6.32.

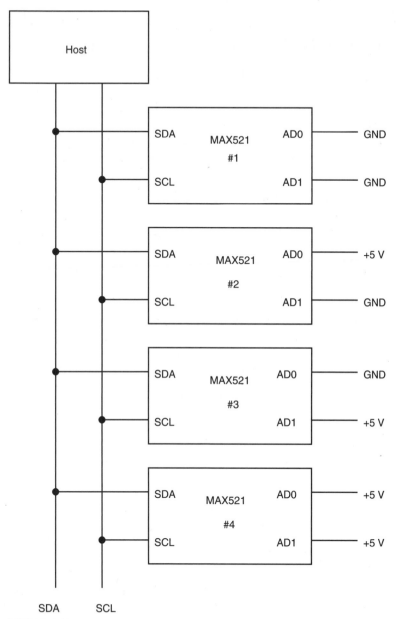

Figure 6.28 Connecting multiple MAX521s on a single bus.

6.9.2 DOT MATRIX DISPLAYS

Dot matrix displays are the best in terms of the type of information that can be displayed, including graphics. Dot matrix displays are necessarily multiplexed displays. Figure 6.33 illustrates an interface circuit for a 5-by-7 dot matrix display. Figure 6.34 is a circuit diagram for a AT90S2313 controlled dot matrix display. The display is arranged as five columns of seven LEDs each. Each column is refreshed at a rate of 40 Hz at 4.00-MHz

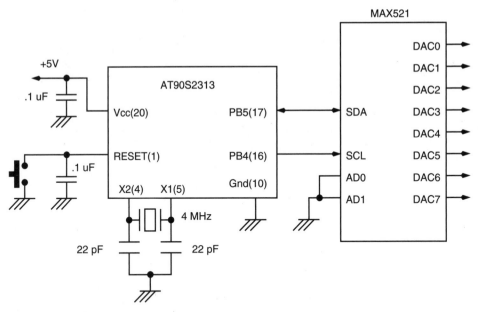

Figure 6.29 Connecting AT90S2313 AVR processor to MAX521 DAC.

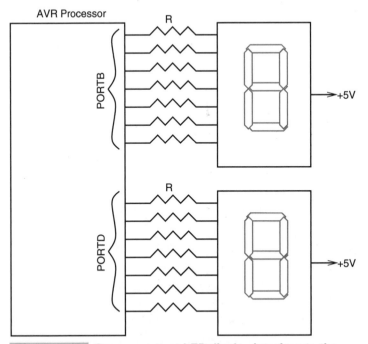

Figure 6.30 Seven-segment LED display interface to the AVR processor.

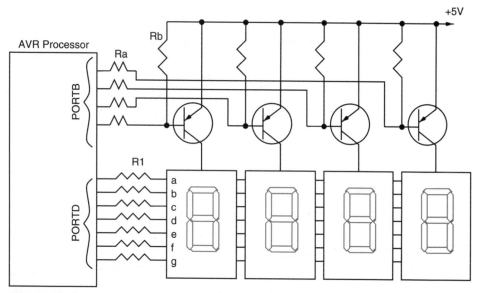

Figure 6.31 A multiplexed seven-segment LED display interface to the AVR processor.

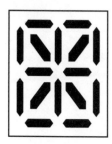

Figure 6.32 An alphanumeric LED display.

clock frequency. The actual clock frequency used for the circuit is 3.58 MHz and so the refresh rate is about 36 Hz.

This is an illustrative circuit. The code just waits for a key to be pressed, and at each key press it displays a new number or a new alphabet in a sequence.

Since there are five columns of LEDs, the duty cycle of current flowing in each column is 20%, and so to maintain the same average current (of 4 mA) the peak current is increased five times to 20 mA. The value of the current-limiting resistor is hence chosen to be 150 ohms. Figure 6.35 illustrates the test board for the 5 × 7 dot matrix display. Code for this circuit is available in the code directory in the file 5x7disp.asm.

6.10   Interfacing LCD Displays

LCD displays are very useful for displaying user information and communication. LCD displays are available in various formats. Most common are 1x16, i.e., 1 line with 16 alphanumeric characters. Other formats are 2x16, 1x40, 2x40, 4x16, etc.

The LCD displays have the following format:

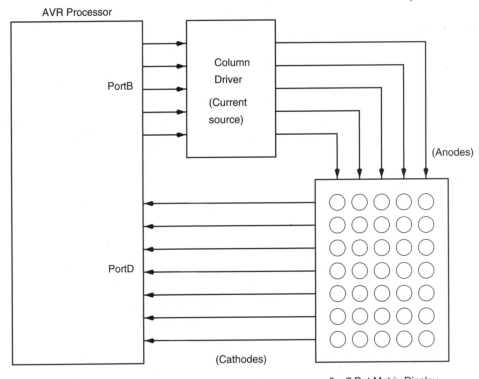

Figure 6.33 Block diagram for a 5-x-7 dot matrix display to AVR interface.

PIN	SYMBOL	I/O	FUNCTION
1	Vss	-	Power supply Gnd
2	Vcc	-	Power supply +5V
3	Vdd	-	Contrast adjust
4	RS	I	0 = Instruction input
			1 = Data input
5	R/W	I	0 = Write to LCD
			1 = Read from LCD
6	E	I	Enable signal
7	DB0	I/O	Data bit line 0 (LSB)
8	DB1	I/O	Data bit line 1
9	DB2	I/O	Data bit line 2
10	DB3	I/O	Data bit line 3
11	DB4	I/O	Data bit line 4

12	DB5	I/O	Data bit line 5
13	DB6	I/O	Data bit line 6
14	DB7	I/O	Data bit line 7 (MSB)

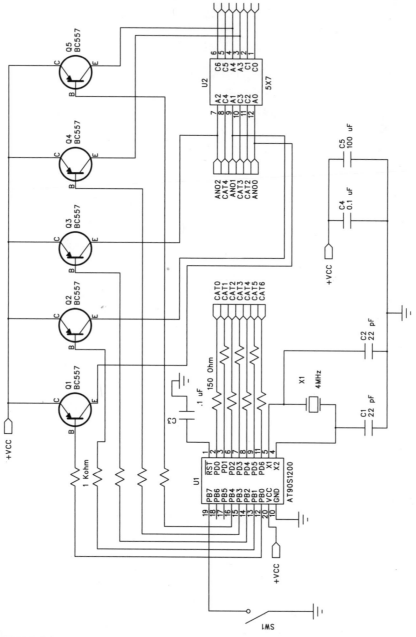

Figure 6.34 Circuit schematic for a 5-x-7 dot matrix display interface.

Figure 6.35 A 5-x-7 dot matrix display test board photograph.

The LCD modules have an 8-bit interface. Besides the 8-bit data bus, the interface has a few other control lines. The default data transfer between the LCD module and an external device is 8 bits, however it is possible to communicate with the LCD module using only four of the eight data lines. Figure 6.36 illustrates the character codes for the LCD, and Figure 6.37 shows how to interface a 2-x-16 line LCD module to an AT90S2313 processor. The R/W line is connected to ground and hence the processor cannot read any status information from the LCD module, but can only write data to the LCD. The source code for the LCD interface example is available on the CD in the code directory in the file my_lcd.asm.

6.11   Driving Relays with AVR

The ULN2003A are high-voltage, high-current darlington arrays containing seven open collector darlington pairs with common emitters. Each of the seven channels can handle 500 mA of sustained current with peaks of 600 mA. Each of the channels has a suppression diode that can be used while driving inductive loads (such as relays) as freewheeling diodes.

The ULN2003A input is TTL compatible. Typical uses of these drivers include driving solenoids, relays, DC motors, LED displays, thermal print heads, etc.

The IC is available in a 16-pin DIP package and other packages. The outputs of the drivers can also be paralleled for higher currents, though this may require a suitable load-sharing mechanism.

Figure 6.38 shows the block diagram of the ULN2003A darlington array driver IC. For each of the drivers, there is a diode with the anode connected to the output and the cathode connected to a common point for all the seven diodes. The outputs are open-collector, which means that external load is connected between the power supply and the output of the driver. The power supply can be any positive voltage less than +50 V as specified by the data sheets. The load value should be such that it needs sustained currents less than 500 mA and peak currents less than 600 mA per driver.

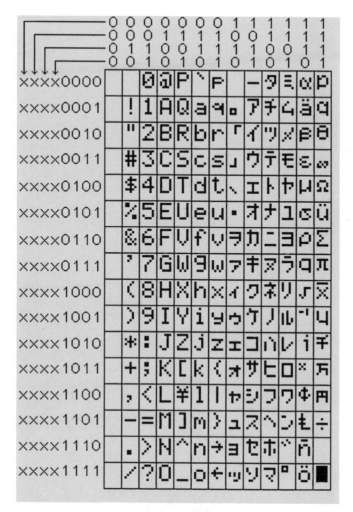

LCD Character Codes

Figure 6.36 LCD character codes.

The following diagram shows how these drivers are used to drive relay coils. Figure 6.39 shows three relays being driven by the outputs of three drivers from the ULN2003A IC. One end of the relay coil is connected to the output of the driver and the other end is connected to the +ve supply voltage. The value of this voltage will depend upon the relay coil voltage ratings. The diode common point is also connected to the +ve supply voltage. The inputs to the ULN2003A IC is TTL voltages, say the output of the port pins of the AVR, for example. With this arrangement, the port signals could be used to control each of the relays.

The relay terminals labeled NC (Normally Closed), common, and NO (Normally Open) could be used to switch whatever voltage that may need to be switched. Typically, the relay terminals are used to switch the main supply (220 V AC or 115 V AC as may be the case) to the required load (a heater or a lamp, etc.), but, of course, it may be used to switch any voltage (AC or DC) as long as the relay contact can handle the voltage and the current.

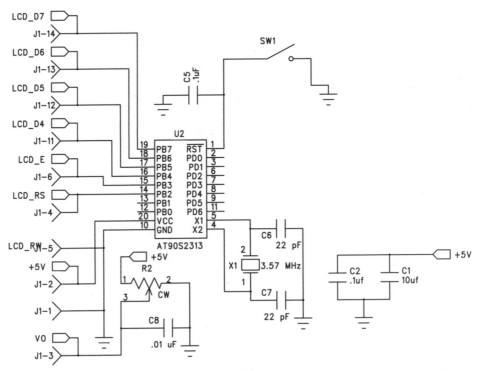

Figure 6.37 Circuit schematic for an AT90S2313 processor interface to a 2-line, 16-character LCD.

6.12 Stepper Motor Interface for the AVR

Figure 6.40 illustrates a very popular stepper motor sequencer and driver interface to the AVR processor. The L297/298 sequencer and driver is made by SGS Thomson (us.st.com).

L297 is a stepper motor controller IC that generates four-phase drive signals for two-phase bipolar and four-phase unipolar step motors in microcontroller-controlled applications. The motor can be driven in half step, normal, and other modes, and on-chip PWM chopper circuits permit switch-mode control of the current in the windings. The IC only requires a mode input, a clock input, and a direction input for its operation. This greatly reduces the software burden of the microcontroller.

To drive the stepper or DC motors, a matching driver IC such as the L298 is used. L298 is a dual full-bridge driver. It can be used with power supply voltages up to 48 V and total DC current up to 4 A.

For a larger drive, L2603 from SGS Thomson can be used instead of the L298. Figure 6.41 illustrates the circuit schematic of L297 and L298, which can be used with an AVR processor.

When moving motors, it is always advisable to gently increase the speed of the motors rather than operate the motor at a fixed speed right at the beginning. The motor is started slowly at a minimum speed and then gradually, the speed is increased till it reaches the

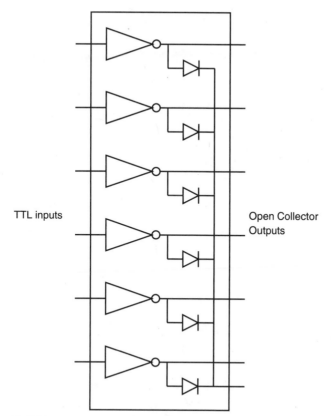

Figure 6.38 ULN2003A darlington array.

maximum operable speed. To bring the motor to a halt, the speed is gradually decreased before stopping. Figure 6.42 illustrates the motor speed ramping.

6.13   Interfacing to a Serial EEPROM

Serial EEPROMs are getting very popular for a variety of reasons. You can store up to 64 Kbytes of data in small 8-pin DIP package. The communication takes only two signals, since most serial EEPROMs are available with IIC interface. These EEPROMs can be written 100,000 times or even more. Even though the data write takes 10 ms, by writing an entire page at a time, the average write rate can be improved. Writing a page of data into the EEPROM also takes almost the same time as writing a byte. The page size can vary between 16 bytes for smaller-capacity EEPROMs to 128 bytes for the larger 64-Kbyte capacity EEPROMs. Thus data transfer speeds can be improved by writing in a burst mode.

Figure 6.47 (appearing later in this chapter) illustrates the circuit schematic for the AVR to EEPROM interface. A MAX232 RS-232 line translator has been connected so that the user can read and write data to the EEPROM.

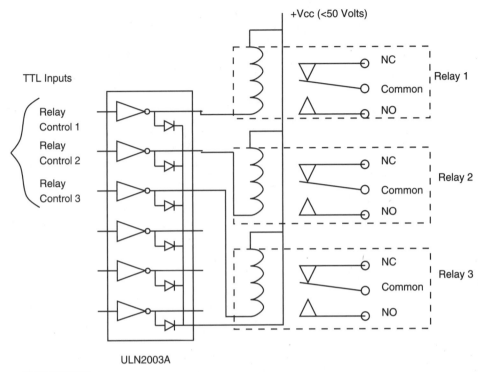

Figure 6.39 ULN2003A drivers used to drive inductive loads.

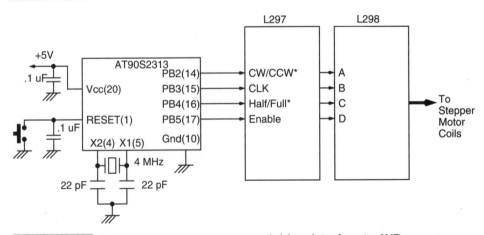

Figure 6.40 A stepper motor sequencer and driver interface to AVR.

The EEPROM has a write protect pin (WP) that can be connected to +5 V to disable any write to the EEPROM. For our use, we have connected it to ground so that we can write data to the EEPROM.

The EEPROM we have chosen is AT24C512 from Atmel. It has a capacity of 64 Kbytes. The EEPROM has two device address lines that allow up to four such EEPROM chips to be connected to the same IIC bus (Figure 6.43).

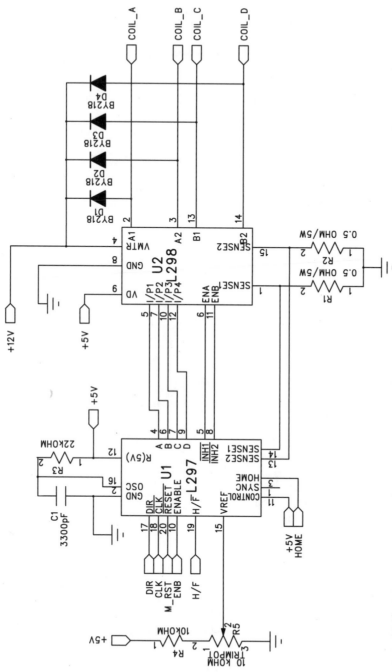

Figure 6.41 Circuit schematic for a stepper motor sequencer and driver for the AVR processors.

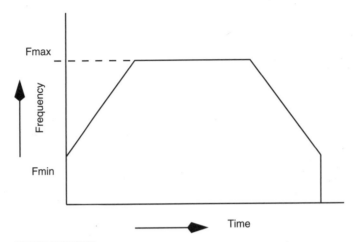

Figure 6.42 Ramping the stepper motor speed.

7	6	5	4	3	2	1	0
1	0	1	0	0	A1	A0	R/W*

Figure 6.43 EEPROM device address.

The EEPROM can be read and written in the following ways:

1. Byte Write: Figure 6.44 illustrates a byte write to the EEPROM. Following the Start sequence, the device address is transmitted with the R/W bit reset to "0", followed by the address of the location to which the byte is to be written to. Since the capacity of the EEPROM is 64 Kbytes, a 16-bit address split as MSB address and LSB address is sent next in that order. In the end, the data byte to be written is sent. The sequence is terminated with a Stop sequence.

2. Page Write: Same sequence as the byte write, except that multiple data bytes that are to be written are transmitted before the Stop sequence is issued.

3. Current Address Read: The Start sequence is issued followed by the device address (R/W bit is set to "1") and the data byte from the EEPROM is received. Figure 6.45 illustrates the transfer.

4. Random and Sequential Read: This requires a dummy write sequence to precede the actual read. The purpose is to provide the address from where to read the data. The Start sequence is issued followed by a device address that includes the R/W bit reset to "0" to indicate write. Then follows the MSB address and LSB address, and after that a new Start sequence is issued, followed by the device address sequence again with the R/W bit set to "1" to indicate the read operation. After this, the device provides the data from the required location and terminated with a Stop sequence. Since the initial write sequence is not terminated with a Stop sequence, the write to the location is not performed; instead, only the address gets changed. Figure 6.46 illustrates the data transfer

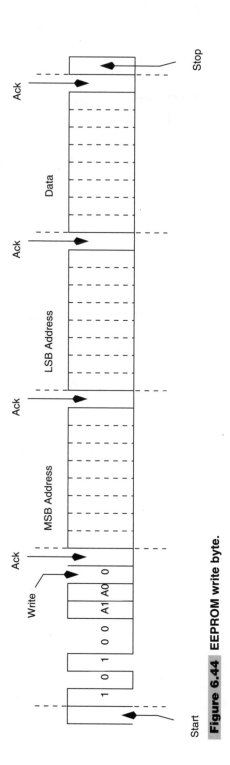

Figure 6.44 EEPROM write byte.

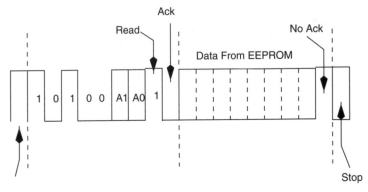

Start

Figure 6.45 EEPROM current address read.

sequence. If an Ack is generated by the processor before the Stop sequence, then data byte from the next location is received, leading to sequential read operation. The last data read must be terminated with a No Ack before the Stop sequence.

The sample program to work with the circuit schematic in Figure 6.47 is available in the code directory in the file ep1byt.asm, which can read and write one byte at a time at the address you specify, and ep2byt.asm, which can read and write two bytes at a time at the specified address.

6.14 Interfacing to a Real Time Clock (RTC)

RTCs are useful devices as timekeepers in embedded systems. Many serial communication format RTCs in 8-pin DIP package with a host of features are available. We will interface DS1302 RTC from Dallas to AVR processors.

This RTC can trickle charge an external standby NiCd battery. It contains 31 bytes of SRAM. The RTC has a simple three-wire interface to a microprocessor. The RTC provides seconds, minutes, hours, day, date, month, and year information. The RTC can operate in a 12-hour format with an AM/PM indicator or a 24-hour format.

The RTC has a synchronous serial communication interface. Only three wires are required to communicate with a processor such as the AVR.

Figure 6.48 illustrates a block diagram on an AVR interface to the DS1302 RTC.

Figure 6.49 illustrates the circuit schematic. MAX232 has been added so that the user can communicate with the processor and read and write to the RTC. The code for this project is available in the code directory in the file rtc_ex.asm.

The data sheet is available here:

http://www.dalsemi.com/DocControl/PDFs/1302.pdf

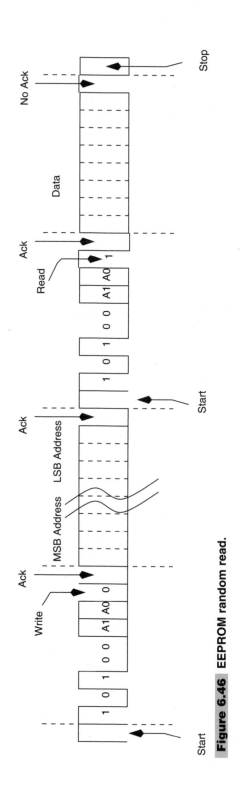

Figure 6.46 EEPROM random read.

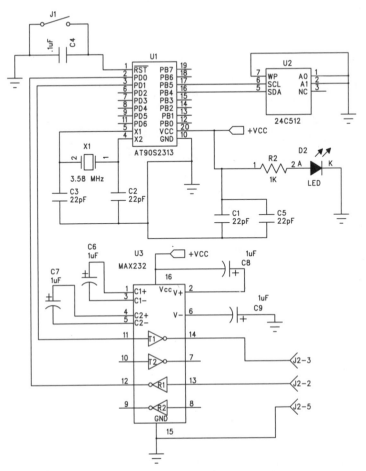

Figure 6.47 Circuit schematic for an AT90S2313 processor interface to a serial EEPROM.

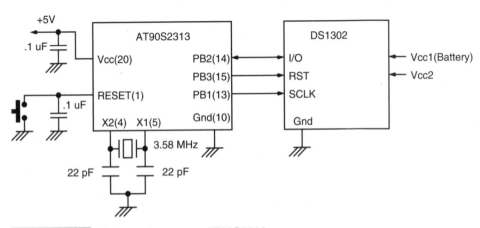

Figure 6.48 RTC interface to an AT90S2313.

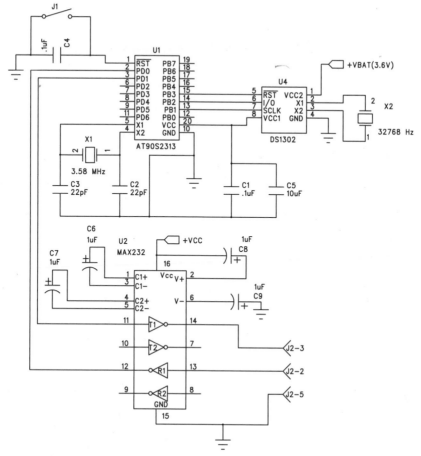

Figure 6.49 Circuit schematic for an AT90S2313 processor interface to an RTC.

6.15   Accessing a Constants Table

Most AVR processors have on-chip flash program memory as well as some amount of EEPROM. Both these memories can be used to store constants. If the constants are stored in the EEPROM, they can even be modified, while the constants stored in the flash program memory cannot be changed expect while programming the flash program memory of the chip.

The constants are stored in the program memory either at a predefined address (using the appropriate origin assembler statement such as .org for the Atmel's AVR assembler) or by identifying the start of the constants table with a label, as in the example below.

```
msg1:  .db "Honk! Honk!    Honk! Honk!    "
```

To access an individual element of the table, say the tenth from the start of the table, the following program is used.

```
        ldi ZH, high(msg1*2)   ;init the pointer register
        ldi ZL, low(msg1*2)
        adiw ZL, 10            ;add an offset to the pointer
more1:  lpm                    ;read program memory.
                               ;data is now available in register R0
```

Similarly, constants and tables can be stored in the EEPROM memory. The assembler program must contain the ".eseg" directive to instruct the assembler to locate the following data in the EEPROM memory map.

```
.eseg
org 0
;Start of the message
morse_msg:
.db 2                          ;C
.db 16                         ;Q
```

Accessing the EEPROM is done as follows:

```
        ldi ZL, low(morse_msg)
eep_notrdy:
        sbic EECR,1            ;skip if EEWE clear
        rjmp eep_notrdy        ;Waits until EEPROM ready
read:
        out EEAR, ZL           ;output address low
        sbi EECR, 0            ;set EERE (Read-strobe)
        nop
        nop
        in R18, EEDR           ;inputs data
```

6.16   Arbitrary Waveform Generation

Generating digital waveforms for various applications is often required, either as a part of a design requirement or as a test pattern generator. Multichannel digital waveform generators are extremely expensive pieces of instruments. Often, you can use a digital circuit to provide a limited functionality of this expensive instrument.

The AVR, with its extremely fast program execution, is quite capable of generating fast, multichannel digital waveforms. An example of what an arbitrary digital waveform might look like is illustrated in Figure 6.50. The required waveform is drawn on a sheet of paper and then encoded as numbers as illustrated in figure wave1. These numbers are then put in a constants table in a program. The waveform generation program outputs the values of the table onto a port which provides the waveform outputs.

Figure 6.51 illustrates one of the waveform patterns being generated by an AT90S8515 processor. The waveform generator code is available on the CD in the code directory in the file wave1.asm.

6.17   A Switch-Case Implementation

The Switch statement is a popular statement used extensively in C. It is essentially a chain of if/else statements. The following code illustrates how a switch-case structure can be implemented on the AVR.

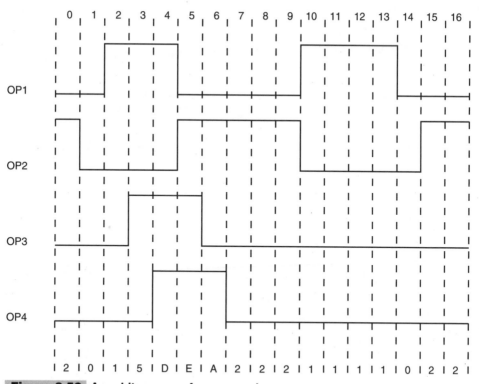

Figure 6.50 An arbitrary waveform example.

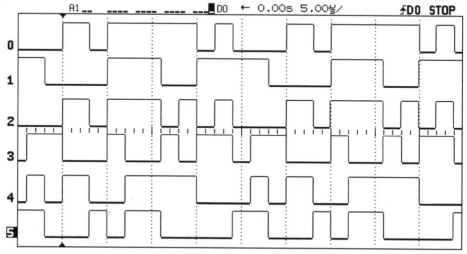

Figure 6.51 An arbitrary waveform generated by the AVR processor and captured on a logic analyzer.

```
            .equ option1='A'
            .equ option2='B'
            .equ option3='C'
            .equ option4='D'
            .equ option5='E'
            .equ option6='F'
            .def rreg=r18
            ;************************************************
            ;Subroutine to implement a switch-case statememt
            ;************************************************
sub_case:   rcall get_byte        ;get an argument
begin_case: cpi rreg, option1     ;check if argu=option1
            brne chk2             ;else compare with option2
            rcall sub_opt1        ;if yes execute subroutine
                                  ;sub_opt1
chk2:       cpi rreg, option2     ;check if argu=option2
            brne chk3             ;else compare with option3
            rcall sub_opt2        ;if yes execute subroutine
                                  ;sub_opt2
chk3:       cpi rreg, option3     ;check if argu=option3
            brne chk4             ;else compare with option4
            rcall sub_opt3        ;if yes execute subroutine
                                  ;sub_opt3
chk4:       cpi rreg, option4     ;check if argu=option4
            brne chk5             ;else compare with option5
            rcall sub_opt4        ;if yes execute subroutine
                                  ;sub_opt4
chk5:       cpi rreg, option5     ;check if argu=option5
            brne chk6             ;else compare with option6
            rcall sub_opt5        ;if yes execute subroutine
                                  ;sub_opt5
chk6:       cpi rreg, option6     ;check if argu=option6
            brne chk_default      ;nothing matches.
            rcall sub_opt6        ;if yes execute subroutine
                                  ;sub_opt6
            ret
chk_default: rcall sub_default    ;else execute a default
                                  ;subroutine
            ret
```

6.18   Implementing a Finite State Machine

A finite state machine (FSM) is a formal concept with many applications. In general, a finite state machine is a device which has some inputs and some outputs. It stores the state of the machine, and depending upon the inputs, changes the state value. The outputs depend upon the state value. FSM has:

1. A set of finite states.
2. Inputs.
3. A transition function for each state.
4. Outputs.

FSM has great applications in pattern recognition, vending machine applications, etc. You can create a traffic light controller modeled on the FSM concept.

The first step in modeling the requirement as an FSM is to create a state transition table and an output table as illustrated in Tables 6.4 and 6.5. Another way is to create a bubble diagram description as illustrated in Figure 6.52. Some of the transitions from one state are conditional, while many others are unconditional. It is valid to have an FSM with all unconditional transitions. An ordinary binary counter is an example of an FSM with unconditional state transitions. The counter just hops from one state to the next and then resets to the starting state. The minimum time for which the FSM remans in a state is determined by the clock period of the system. In a microcontroller-based system, you could put appropriate delay routines between each transition.

The state of the FSM is maintained with the help of a variable (in a digital circuit, with the help of a register). It is important that all the possible states of the register are accounted for and, importantly, after power on, the register (or the variable) must be initialized with a default state, otherwise the FSM will not work at all. A test code for implementing a finite state machine is available in the code directory in the file fsm.asm.

TABLE 6-4   STATE TRANSITION

CURRENT STATE	NEXT STATE (NS)
S0 X = 0 and Y = 0,	NS = S1
SO X = 1, NS = S0	
S1 X = 1 and Y = 0,	NS = S2
S1 X=1 and Y = 1,	NS = S3
S2	NS = S3
S3	NS = S4
S4	X = 1, NS = S4
S4	X = 0, NS = S0

TABLE 6-5   STATE OUTPUT

STATE OUTPUTS
SO A = 0, B = 23
S1 A = 1, B = 23
S2 A = 1, B = 32
S3 A = 0, B = 22
S4 A = 1, B = 44

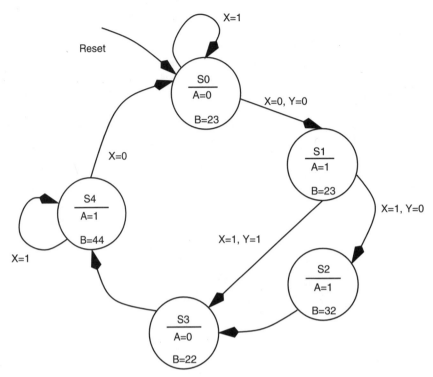

Figure 6.52 A bubble diagram description of a state machine.

6.19   Generating Random Numbers

Many applications such as toys and test pattern generators require random numbers. While it is almost impossible to generate a truly random number, one can approximate with a pseudorandom number.

One popular way to generate a pseudorandom number is to read the contents of a free-running counter. This is a simple scheme and can be easily implemented on the AVR controller with no extra hardware. Timer0 (or Timer1) is clocked at a certain clock frequency derived out of the system clock. Then, to get a random number, the Timer0 (or Timer1) register (TCNT0 or TCNT1) is read and what you get is a pseudorandom number. This scheme is used in the Dice project chapter.

Another way to generate a pseudorandom number is to use the concept of Linear Feedback Shift Register (LFSR). LFSRs are ordinary shift registers with some outputs (called taps) feeding the input (see Figure 16-3 in Chapter 16). LFSRs have an interesting property that if the feedback taps are chosen carefully, then outputs cycle through $2^n - 1$ sequences, for an n-bit LFSR. The sequence then repeats after $2^n - 1$ instances. If the output sequences are observed, they appear to be random. An 8-bit LFSR is illustrated in Figure 16.3. An 8-bit LFSR will have a sequence length of 255. Similarly, a 16-bit LFSR would have a length of 65535 and so on.

The LFSR can be easily implemented on the AVR controller. The LFSR must be initialized with a nonzero seed value. After the LFSR is initialized, it is clocked by shifting the values to the left and loading a new bit into the bit0 of the shift register. The new bit that is loaded into the bit0 of the shift register is calculated by XORing the bits at the selected taps of the LFSR. All of these operations can be implemented using the AVR instructions. A working example of an 8-bit LFSR implemented on the AVR controller is presented in the electronic lock project chapter.

BITS	SEQUENCE LENGTH	TAPS
9	511	3,8
10	1023	2,9
11	2047	1,10
12	4095	0,3,5,11
13	8191	0,2,3,12
14	16,383	0,2,4,13
15	32,767	0,14
16	65535	1,2,4,15
17	131,071	2,16
18	262,143	6,17
19	524,287	0,1,4,18
20	1,048,575	2,19

7

COMMUNICATION LINKS
FOR THE AVR PROCESSOR

7.1 Introduction

No man is an island, and neither is a microcontroller application. Except for those that are completely self-contained in a single chip microcontroller, your microcontroller-based application will need to communicate with an external device, be it an additional peripheral device, a host PC, or another microcontroller-based application. Figure 7.1 illustrates the idea.

In the simplest of cases, the AVR device communicates to a peripheral device such as a memory device, or a digital port or an ADC. These devices may be part of a single application. In another case, your AVR device may want to communicate with a host PC for transferring data, while in another case, it may want to communicate to multiple devices all interconnected to each other using a bus configuration.

The communication between an AVR and an external device or devices is essentially of two types: point-to-point communication or a bus-based communication. A point-to-point communication connects two devices, while a bus offers multiple devices to share the same physical connection lines.

Whether to go for a point-to-point communication or a bus-based communication is solely determined by need. Bus-based communication seems attractive, as a single communication link allows multiple devices to communicate, but this scheme is not without complexities—for example to regulate communication traffic on the shared lines.

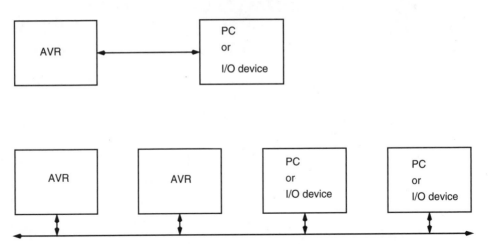

FIGURE 7.1 Communication link for AVR processor. The figure illustrates the processor in a point-to-point communication link to another device as well as a link with a bus configuration with multiple devices connected onto the bus.

While as a designer you are free to design your own communication protocol, both at a hardware and software level, it is best to consider one of the many standard communication schemes. The added advantage is that the AVR has many of these schemes available as on-chip peripherals.

7.2   RS-232 Link

RS-232 communication is by far the most common communicating mode that the AVR can utilize. RS-232 is an asynchronous serial transfer mechanism. This bit-serial transmission method can be split up in two parts: the way the original byte data is split up serially for transmission and the way this serial data is physically transmitted over wires.

The way the data is split serially is not unique to RS-232. Other protocols like RS-422, RS-423, and RS-485 also use this method. These asynchronous serial methods differ in the way the serial data is transmitted over the physical wires.

Figure 7.2 shows how the original data is reorganized with a start bit added at the beginning of the data transmission and at the end, an optional parity bit, and up to two stop bits. Also, the order of bit transmission is LSB of the data byte first and MSB of the data byte last.

Once the data to be transmitted is lined up as in Figure 7.2, it is time to consider how is it physically transmitted over wires, what the voltage levels are, and what the duration of each transmission is.

The transmitter and receiver of data use a fixed data rate, called the bit rate. Most common bit rates are 300, 600, 1200, 1800, 2000, 2400, 4800, 9600, and 19200 bits per second. Bit rate is the time for which one bit (out of the 10 or so bits) is available at the output (or input).

The bit data to be transmitted is converted to RS-232 standard voltage level before putting it on the wires. The legal voltage limits are illustrated in Figure 7.3.

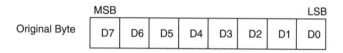

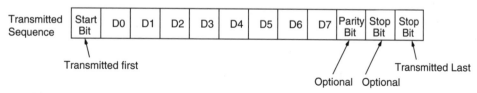

FIGURE 7.2 How the data is reorganized and extra bit attachments added to the original bit sequence in asynchronous serial data transmission.

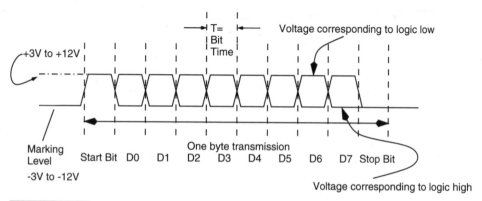

FIGURE 7.3 Voltage levels on the RS-232 serial transmission. The wave form is illustrated without any parity bit and one stop bit.

While Figure 7.3 illustrates the voltage levels on an RS-232-compatible line, it is possible to connect to AVR devices without the use of RS-232 level converters. Though in that case, the voltage transmission won't be RS-232 compatible (Figure 7.4). It is also advisable to use this configuration with very short wire links (few inches). For true RS-232 compatibility, it is necessary to use one of the many RS-232 line driver/receiver chips illustrated in Table 7.1. Using these driver chips, one can safely connect maximum cable lengths up to 15 meters. The table is not exhaustive, though.

The RS-232 standard is a complex standard and has many signal lines. Originally, the standard specifies a 25-pin "D" type connector as well as a 9-pin "D" type connector with all the signals. Of these signals, it takes only TxD, RxD, and Gnd to put together a simple duplex RS-232 communication link. Most controller chips (including some of the processors in the AVR family) are also equipped with a bare-minimum serial interface which includes only TxD and RxD signal lines. Other signal lines of the interface are called handshake lines, which are used by two devices to receive data correctly and in a regulated manner.

For smooth data transfer between two devices, the RS-232 standard uses the concept of DTE (Data Terminal Equipment) and DCE (Data Communication Equipment). The pin

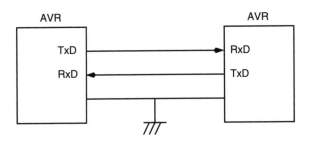

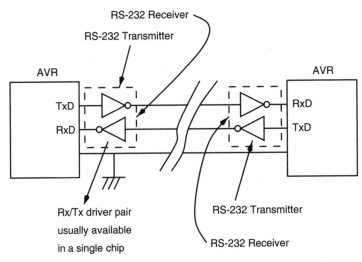

FIGURE 7.4 Connecting an AVR device to another AVR or any other serial device.

outs and function names referred to in Table 7.2 are with respect to the DTE. The DCE has the same connector pinout, except the direction of the pins reverses. Thus pin 2 of a 9-pin connector on a PC configured as DTE is RxD as an input, while on a modem with a 9-pin connector seen as a DCE, this pin is still called RxD, except it is an output.

While we won't be using most of the handshake lines in the rest of the book for any real handshake purpose, these handshake signal lines can be used for another interesting functionality: to provide power to our AVR circuits! While connecting AVR circuits to full function RS-232 ports on a PC, we can use signal lines RTS and DTR to provide supply voltage for the microcontroller.

The AVR family has many processors equipped with on-chip serial port functionality. With a suitable line driver and receiver component (refer again to Table 7.1), it is possible to create a RS-232 link.

7.3   RS-422/423 Link

The RS-422/423 communication differs from the RS-232 in only the way the bit logic levels are translated into line voltages, the maximum possible data rates, and the length of the

TABLE 7-1   SOME RS-232 LINE DRIVER AND RECEIVER ICS

MANUFACTURER	PART NUMBERS	COMMENTS
Maxim	MAX212	low power, +3 V operations, 3 Tx and 5 Rx channels, up to 120 Kbps.
Maxim	MAX232	+5 V operation 2 Rx, 2 Tx channels
Dallas	DS232	MAX232 compatible
Dallas	DS275	Line powered, half duplex (i.e., either transmit or receive at a time)
Intersil	HIN203	No capacitors required!, up to 120 kbps
Analog dev.	ADM232	MAX232 compatible

TABLE 7-2   RS-232 SIGNALS AND CONNECTOR PINOUTS

25-PIN	9-PIN	NAME AND FUNCTION	DIRECTION
1	—	Gnd —	
2	3	TxD, Transmit Data	Out
3	2	RxD, Receive Data	In
4	7	RTS, Request To Send	Out
5	8	CTS, Clear To Send	In
6	6	DSR, Data Set Ready	In
7	5	Gnd —	
8	1	DCD, Data Carrier Detect	In
20	4	DTR, Data Terminal Ready	Out
22	9	RI, Ring Indicator	In

cable. RS-422 uses a differential, balanced protocol for communication and is good for up to a maximum of 10 Mbits/s and up to a distance of about 1000 meters (not at the maximum speed though).

Figure 7.5 illustrates a RS-422 link. Many manufacturers make RS-422 drivers, and MAXIM's MAX488/490 ICs are quite good (Figure 7.6).

7.4   RS-485 Link

The RS-485 communication also differs from the RS-232 in the way the bit logic levels are translated into line voltages, the maximum possible data rates, and the length of the cable and the direction of traffic. RS-485 is a communication bus and allows multiple devices to communicate on the link. As a consequence, it is a half-duplex link, i.e., at any given time a device

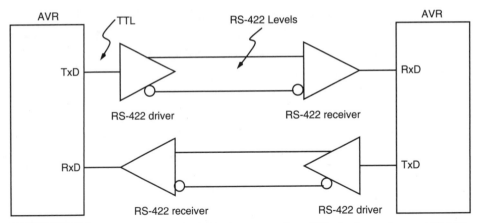

FIGURE 7.5 Connecting an AVR device to another AVR device using an RS-422 link.

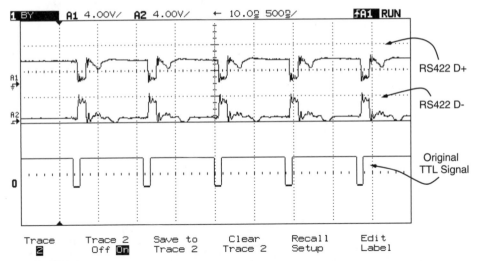

FIGURE 7.6 Original data and the corresponding differential outputs of an RS422 driver.

on the bus can either receive data or transmit data. Figure 7.7 illustrates the way an RS-485 link for multiple AVR devices could be configured. Since at a given time, only one of the AVR processors can transmit, the rest of the AVR processors must remain as receivers. To achieve this, the RS-485 driver IC has an enable pin for the transmitter as well as the receiver. These pins are connected to other port pins of the respective AVR processor, so that under software control, the AVR can decide whether to transmit or to receive data.

This puts more software overheads on the system. Also, the designer must decide beforehand which of the AVR processors is going to be the master processor at power on. After the power is applied (or after system reset), only this AVR device enables its transmitter and sends out data on the bus. The reset of the processors enables their respective receiver and receive data from the bus. The control of the bus (i.e., which AVR device gets to transmit) is subsequently transferred in software between these AVR processors as

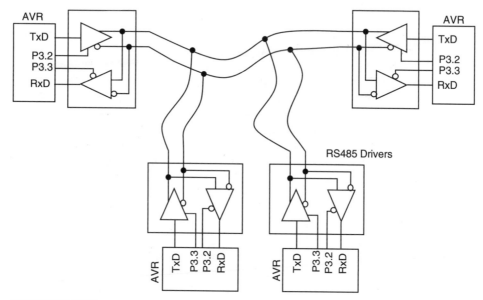

FIGURE 7.7 Connecting multiple AVR devices on a RS-485 bus.

desired. However, it has to be ensured by the software on all the devices that one and only one AVR transmits on the bus at a given time. For this purpose, it is usual to allocate unique device numbers to the various processors.

The RS-485 data speeds can be up to a maximum of 10 Mbits/s and up to a maximum distance of 1000 meters (but not at maximum speed). It also allows up to 32 drivers and receivers on the bus. MAXIM makes RS-485 drivers such as MAX485, which is a drop-in replacement for the immensely popular LTC485 from Linear Technology.

7.5 SPI and MICROWIRE Bus

The Serial Peripheral Interface (SPI) circuit is a synchronous serial data link that is standard across many Motorola microprocessors and other peripheral chips. It has become so popular that many other controllers support it, including the AVR processors. As the name implies, the SPI is used to allow the controller to communicate with the peripheral devices. SPI peripheral devices from simple shift registers to ADCs, DACs, and memory chips are available. SPI supports a high data rate of up to 3 MHz. Controllers with integrated SPI ports are available to connect to peripheral devices with SPI ports. However, the controller can also use its ordinary I/O lines to mimic an SPI port, though at the cost of data transfer speed.

The SPI port has the following signals:

1. MISO: Serial Data Output signal.
2. MOSI: Serial Data Input Signal.
3. SCK: Serial Clock.
4. Select signal.

Many of the AVR processors are equipped with an SPI port. The lower-end devices that do not have this capability can be used to connect to SPI peripheral devices using the timing diagram illustrated in Figure 7.8. The figure illustrates the data output by the master to the slave on the MOSI line and the validity of this data with respect to the serial clock SCK signal. The other section of this diagram illustrates the way the data is read out of the slave on the MISO line by the master. In the AVR processors, the SCK clock frequency is limited by the system clock frequency and with current processors, it can go up to about 2.5 MHz. It should be noted that the data transfer in an SPI system is in multiples of 8 bits at a time.

The MICROWIRE is a communication protocol similar to the SPI, but the data transfer is in chunks of 16 bits at a time. The data transfer speed is also limited to 1 MHz.

7.6 IIC Bus

The IIC or I2C (Inter IC) bus is a popular bus for peripheral expansion and inter IC control. It was designed and propagated by Philips for use in their consumer products.

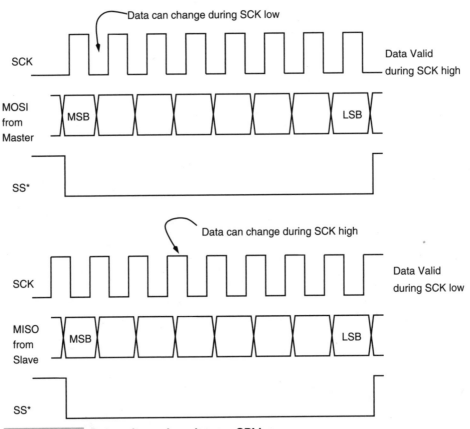

FIGURE 7.8 Data write and read on an SPI bus.

I2C is a two-wire bus interface with communication lines called a Serial Data Line (SDA) and a Serial Clock Line (SCL). Each device connected on the bus is software addressable by a unique address. On the bus there exists a master-slave relationship between the devices. The data transferred between the devices is in multiples of 8 bits. The data transfer rate is 100 Kbit/s in standard modes, 400 Kbit/s in fast mode, and 3.4 Mbit/s in high-speed mode.

Figure 7.9 shows how various components connect to each other on a I2C bus. Of course, these components need to have the IIC bus interface. A complete system usually consists of a microcontroller and other peripheral devices.

The communication on a IIC bus is initiated by the bus master (Figure 7.10). The master generates clock signals (SCL) to permit data transfer. At this time, the device that is addressed is a slave. All communication starts with a START condition and ends with a STOP condition, as illustrated in Figure 7.11. Thus these lines cannot be shared with any other non-IIC capable device except in a very simple case of a fixed master and slave, which is discussed in a later project chapter.

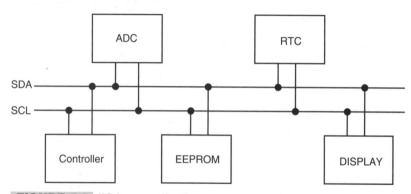

FIGURE 7.9 IIC bus application.

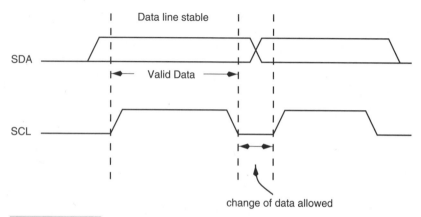

FIGURE 7.10 Bit transfer on an IIC bus.

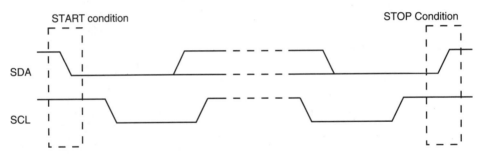

FIGURE 7.11 START and STOP conditions on an IIC bus.

Any number of bytes can be transferred between the master and the slave, depending upon the requirement, and it varies from device to device. But each byte has an acknowledge bit, thus each byte transfer requires nine clock cycles. The receiver must pull the SDA bit down to logic "0" during the acknowledge clock pulse (i.e., SDA is low when the acknowledge clock pulse is high).

The AVR processor is not equipped with an integrated IIC port but can communicate with any IIC interface using ordinary port lines and suitable driver software. Besides, interfacing to many peripherals, the AVR can use IIC expanders to increase I/O capability with a minimum of port lines. Philips offers many such port expanders.

7.7 PC Parallel Port

The Parallel Port is one of the ubiquitous ports on the PC. Originally meant to be used for connecting printers to the PC, its use has proliferated extensively. It is often used to connect external disk drives (the ZIP drive is a common example), network controllers, scanners, data acquisition systems, electronic locks (more on that in a later chapter), etc.

The AVR can be used to control applications that need PC connectivity. We have a complete chapter that shows how the AVR can be connected to the PC parallel port, but in this section, we just look at the features of the parallel port and understand what can be done with this interface.

Figure 7.12 illustrates the inner detail of the parallel port. In reality, the parallel port is not just a single port but a composite of three ports. These ports are:

1. The 8-bit DATA port as an output port.
2. The 4-bit CONTROL port as an output port.
3. The 5-bit STATUS port as an input port.

The port signals are available on a 25-pin "D" type female connector on the PC, as illustrated in Figure 7.12.

Further, the DATA port has two ports, one output port and another input port at the same address. This input port at the DATA port address is really used to read the state of the DATA port pins and is often called the read-back port. Similarly, the output CONTROL

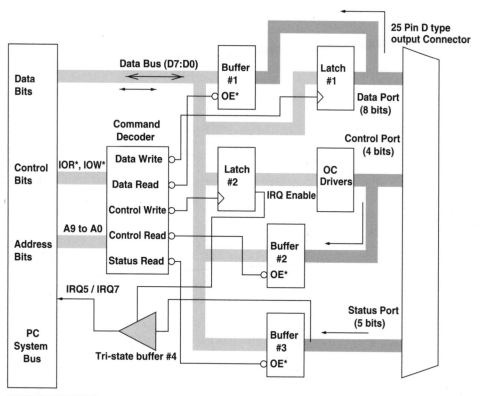

FIGURE 7.12 The details of the PC parallel port.

port also has an input port at the same address and is used to read the status of the CON-TROL port pins. So, in reality the parallel port has five ports.

Figure 7.13 illustrates the detail of the DATA port. The figure shows the internal PC bus connected to an output latch. The outputs of this latch are the DATA port outputs. Similarly, these 8 bits of the data port are also connected to the input of a buffer, which connects to the PC bus. Thus a program running on the PC could output any combination of 8 bits on the DATA port through the output DATA port and, when required, read this combination back through the corresponding input port.

It is easy to access the DATA port either for sending data out or reading the state of the DATA port pins using a C program running on the PC. However, to be able to send data to or read data from any port on the PC, it is essential to know the port address. These port addresses are not completely fixed. Different PCs may have different addresses. Rather than guess the address values, the user can find out the actual port addresses in a PC by reading the contents of certain RAM locations called BIOS data area (the address of these RAM locations is known and fixed!). For the PC parallel port, the RAM address is 0040:0008 (hex) for LPT1, 0040:000a (hex) for LPT2, and 0040:000c (hex) for LPT3. In a particular PC, if the number found in these locations turns out to be zero, then it means that the corresponding LPT does not exist on the PC. A nonzero number on these locations indicates that the corresponding LPT exists. This number is the base address of the parallel port. The port addresses of the various ports are as follows:

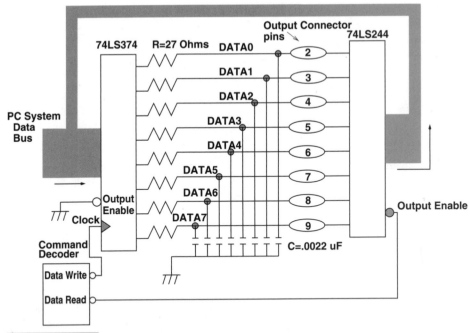

FIGURE 7.13 The DATA port.

1. DATA Port = Base Address
2. STATUS Port = Base Address + 1
3. CONTROL Port = Base Address + 2

I will show some C code to do various things with the parallel port in this section. To read the base address and to calculate the port addresses of the various ports of the parallel port, do the following:

```
int offset, DATAport, STATUSport, CONTROLport, BASEaddress;
offset = 0x08; /*0x08 for LPT1, 0x0a for LPT2, 0x0c for LPT3*/
BASEaddress = peek(0x40, offset); /*read the address from BIOS data
area*/
DATAport = BASEaddress; /* address of DATA port*/
STATUSport = BASEaddress + 1; /*address of STATUS port*/
CONTROLport = BASEaddress + 2; /*address of CONTROL port*/
    To write some value to the DATA port pins, we do the following:
int offset, DATAport, STATUSport, CONTROLport, BASEaddress;
unsigned char data_value;
data_value = 0x55;
/*write a binary sequence 01010101 on the DATA port pins*/
outportb(DATAport, data_value);
    To read the state of the DATA port pins, we do the following:
int offset, DATAport, STATUSport, CONTROLport, BASEaddress;
unsigned char data_pins;
data_pins = inportb(DATAport);
printf(``\nThe DATA pins have the value = %x'', data_pins);
```

Figure 7.14 illustrates the detail of the STATUS port. The figure shows the internal PC bus connected to an input buffer. The inputs of this buffer (some are through inverters) are the STATUS port inputs. The STATUS port has 5 inputs only connected to the upper 5 bits of the 8-bit port. The STATUS port is used to read external data into the PC.

The following code shows how to read external data from the STATUS port and how to press the date to account for the inversions on certain bits.

```
int STATUSport;
unsigned char status_pins;
status_pins=inportb(STATUSport) & 0xf8;
printf(""\nActual STATUS port = %X'', status_pins);
printf(""\nSTATUS port shifted = %X'', status_pins}}3);
printf(""\nSTATUS port shifted and corrected = %X'', 0x10 ^
(status_pins>>3));
```

Figure 7.15 shows the block diagram for the CONTROL port of the parallel port. The CONTROL port has 4 bits which are placed in the lower 4 bits of a byte. Of the 4 bits, 3 of the bits are inverted before being output of the port pins. These pins are also read back through an inverting buffer. An inverter on one of the lines accounts for the signal inversion on the output lines, and if this buffer is read, it provides the bit information that was originally sent to the CONTROL port.

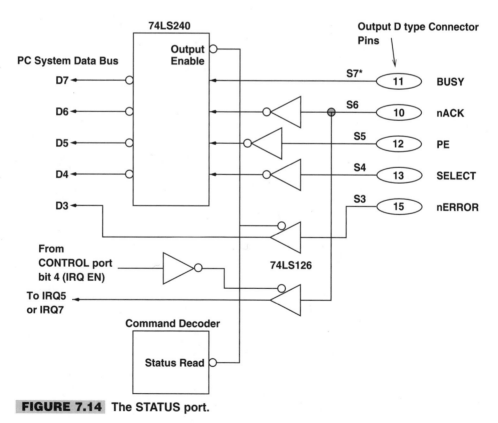

FIGURE 7.14 The STATUS port.

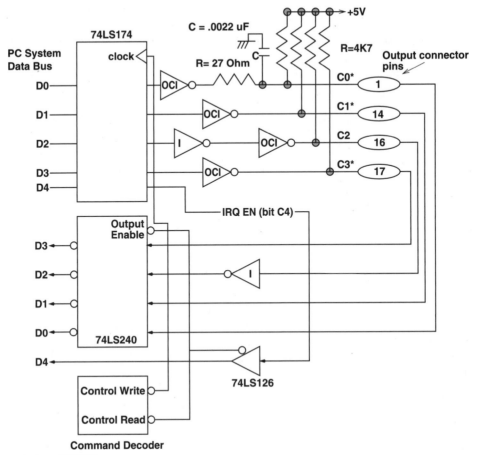

FIGURE 7.15 The CONTROL port.

The data can be sent on the CONTROL port in the same manner as the DATA port (except that the port address is at an offset of two from the base address) and the CONTROL port data is read back in the same manner as the data is read on the STATUS port, again with the offest of two from the base address.

Table 7.3 illustrates the signals, connector pin numbers, and functions of all the parallel port signals. A working example of interfacing an AVR processor to the parallel port is illustrated in a later chapter as a project.

There is a simple arrangement to connect an AVR processor and the parallel port (Figure 7.16). This mechanism allows 4 bits to be transferred between the parallel port and the AVR. We have chosen the CONTROL port, as it can be used both for data output and data input. To output data, appropriate data is written to the CONTROL port and when reading data, first the CONTROL port bits are all set to "1" (logic high), and then the read buffer (at the same address as the CONTROL port) on the CONTROL port is read. The read value (after correction for the inverters) is the value output by the AVR.

One of the STATUS port lines is used by the parallel port as one of the input handshake lines. The output handshake line is by using the DATA port pin as illustrated.

TABLE 7-3   THE SIGNALS OF THE CENTRONICS PARALLEL PRINTER ADAPTER

DB-25	CENTRONICS	REG.	I/O	BIT	NAME	FUNCTION
1	1	Control	Out	C0*	nSTROBE	Active low. Indicates valid data is on the data lines.
2	2	Data	Out	D1	DATA-1	8 data lines
to	to			to	to	Output only in
9	9			D8	DATA-8	older SPP
10	10	Status	In	S6	nACK	A low asserted pulse to indicate that the last character was received.
11	11	Status	In	S7*	BUSY	A high signal asserted by the printer to indicate that it is busy and cannot take data.
12	12	Status	In	S5	PE	Paper empty
13	13	Status	In	S4	SELECT	Asserted high to indicate that the printer is online
14	14	Control	Out	C1*	AUTO FEED	Active low. Instructs the printer to automatically insert a line feed for each carriage return.
15	32	Status	In	S3	nERROR	Signal by printer to the computer to indicate an error condition.
16	31	Control	Out	C2	nINIT	Active low. Used to reset printer
17	36	Control	Out	C3*	nSELECTIN	Active low. Used to indicate to the printer that it is selected
18, 19, 21, 23 to 25, 27, 29 30, 34	25					GROUND

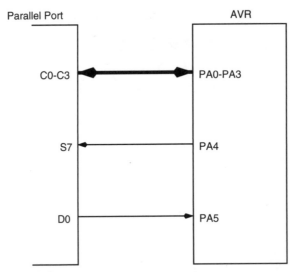

FIGURE 7.16 Connecting an AVR processor to the PC parallel port.

7.8 ISA Bus

The ISA (Industry Standard Adapter) bus is inside the PC used by the processor to communicate with its peripherals. However, it is slowly being replaced by the faster PCI bus. Even so, it is a useful bus to understand and can be used in many applications, as it is easy to build and is fairly fast.

This bus is accessible only on the PC motherboard through a 62-pin female connector. You can connect to it with the help of a PCB with a matching male-type printed connector. The ISA bus has signal lines for data, address, control and power.

Figure 7.17 shows how an AVR processor (even an 8-pin processor can be connected this way) can be connected to the ISA bus. The strobe and busy signals are used for handshaking and synchronizing data transfers between the PC and the AVR processor.

The way the data transfer is synchronized is illustrated in Figure 7.18. A working example of data transfer using handshake signals is illustrated in one of the projects later.

The ISA bus signals are illustrated in Figure 7.19. The signals and their functions are:

1. D0-D7 (I/O): The data bus. The 8-bit data bus is bidirectional and is used for data transfer from and to the adapter cards that fit into the card slots.
2. A0-A19 (O): The address bus has 20 bits and indicates the address of the data transfer between the CPU and other devices or the DMA controller and other devices.
3. IOW* (O): This signal is generated either by the processor or the DMA controller to indicate that data transfer to the addressed destination port is in progress.
4. IOR* (O): This signal, generated by the processor or the DMA controller, indicates that data is read from the addressed port (Figure 7.20).
5. MEMW*(O): This signal, generated by the CPU or the DMA controller indicates that the CPU or the DMA controller wants to write data into the addressed memory location (Figure 7.21).

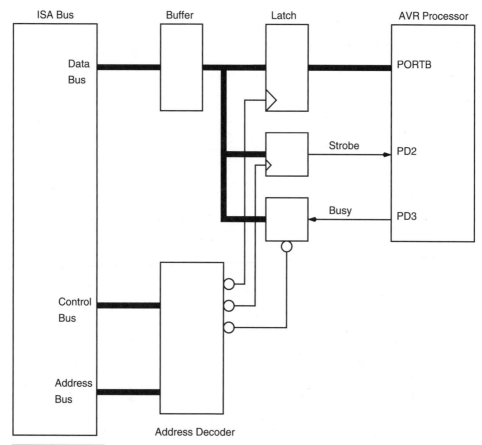

ISA Bus   Buffer   Latch   AVR Processor

Data Bus

PORTB

Strobe   PD2

Busy   PD3

Control Bus

Address Bus

Address Decoder

FIGURE 7.17 ISA bus interface for the AVR.

6. MEMR* (O): This signal, generated by the CPU or the DMA controller, indicates that the CPU or the DMA controller wants to read data from the addressed memory location.

7. RESET DRV (O): This signal provides the reset signal to ports and other devices during power up or during a hardware reset. It is an active high signal.

8. IRQ2-IRQ7 (I): These are the interrupt inputs to the Priority Interrupt Controller (PIC) chip on the motherboard.

9. CLK (O): This is the highest frequency available on the card slot and is three times the OSC frequency.

10. OSC (O): This is the clock signal to which all the IOW* and other strobe signals are referenced. It has a frequency between 4.77 MHz on the original PC to 8 MHz on the new PCs.

11. ALE (O): The is the Address Latch Enable signal. During a transfer from/to the CPU, the CPU places the address on the address lines. The original CPU had the lower 8 address lines multiplexed with the 8 data bits. The ALE signal is used as a demultiplexer signal for the address information. On the system bus, the address and the data bits are already demultiplexed and the ALE signal is only used as sync signal to indicate the beginning of a bus cycle.

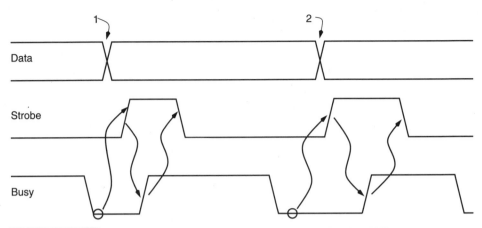

FIGURE 7.18 ISA bus interface data transfer protocol for the AVR.

12. TC (O): This signal is generated by the system DMA controller to indicate that one of the channels has completed the programmed transfer cycles.

13. AEN (O): The AEN signal is generated by the DMA controller to indicate that a DMA cycle is in progress. A DMA cycle could involve a port read and a memory write. However, the port address on the expansion card should not respond to the port read bus cycle if it is not intended. By using the AEN signal, the card circuit can detect if the bus cycle is issued by the CPU or the DMA controller and respond accordingly. A high AEN indicates the bus cycle issued by a DMA controller.

14. I/O CH RDY (I): This signal is used by the card circuit to indicate to the CPU or the DMA controller to insert wait states in the bus cycle. Up to 10 clock cycles can be inserted.

15. I/O CH CK* (I): This signal can be used by the circuit on a plug-in card to indicate an error to the motherboard. A NMI corresponding to INT2 is generated by the motherboard circuit in response to a low I/O CH CK* signal.

16. DRQ1-DRQ3 (I): This is an input signal to the DMA controller on the motherboard. When a port device wants to transfer data to and from the memory, it can use the DMA transfer cycle to do that. The operation of the DMA transfer cycle is controlled by the DMA controller. DRQ1 to DRQ3 are the three inputs to the DMA controller. The system ROM BIOS put DRQ1 at the highest priority and DRQ3 at the lowest at reset. The DMA controller has four channels, but one of them, DRQ0, is used on the motherboard to generate dummy read cycles to refresh dynamic memory.

17. DACK0*-DACK3* (O): These are the four status outputs of the DMA controller that indicate the acceptance of the DRQ request. The DMA transfer cycles begin after the DACK* line is put to "0."

18. Power supply: The motherboard provides $+5$ V, $+12$ V, -5 V, and -12 V voltages to the card slots. The +ve voltages are guaranteed to be within $\pm5\%$ of their nominal values, and the -ve voltages between $\pm10\%$.

7.9   Universal Serial Bus

The USB (Universal Serial Bus) is one the most upcoming and recent interfaces available on the PCs. It is being used to connect all sorts of peripheral devices to the PC, including

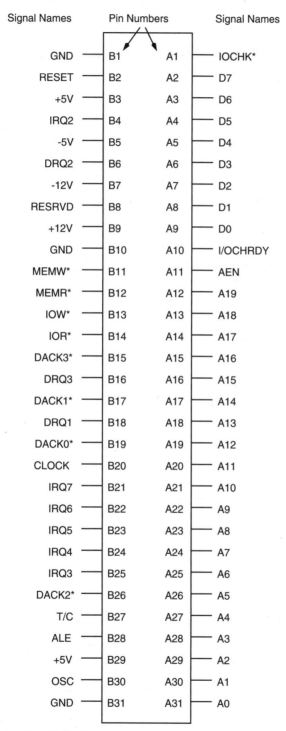

Signal Names	Pin Numbers		Signal Names
GND	B1	A1	IOCHK*
RESET	B2	A2	D7
+5V	B3	A3	D6
IRQ2	B4	A4	D5
-5V	B5	A5	D4
DRQ2	B6	A6	D3
-12V	B7	A7	D2
RESRVD	B8	A8	D1
+12V	B9	A9	D0
GND	B10	A10	I/OCHRDY
MEMW*	B11	A11	AEN
MEMR*	B12	A12	A19
IOW*	B13	A13	A18
IOR*	B14	A14	A17
DACK3*	B15	A15	A16
DRQ3	B16	A16	A15
DACK1*	B17	A17	A14
DRQ1	B18	A18	A13
DACK0*	B19	A19	A12
CLOCK	B20	A20	A11
IRQ7	B21	A21	A10
IRQ6	B22	A22	A9
IRQ5	B23	A23	A8
IRQ4	B24	A24	A7
IRQ3	B25	A25	A6
DACK2*	B26	A26	A5
T/C	B27	A27	A4
ALE	B28	A28	A3
+5V	B29	A29	A2
OSC	B30	A30	A1
GND	B31	A31	A0

FIGURE 7.19 ISA bus signals.

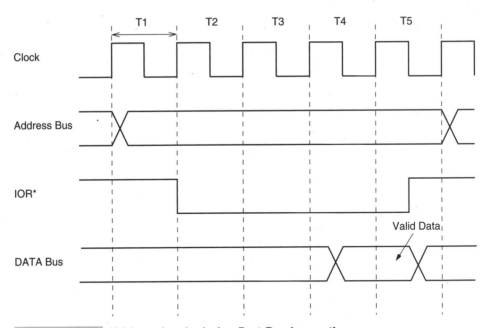

FIGURE 7.20 ISA bus signals during Port Read operation.

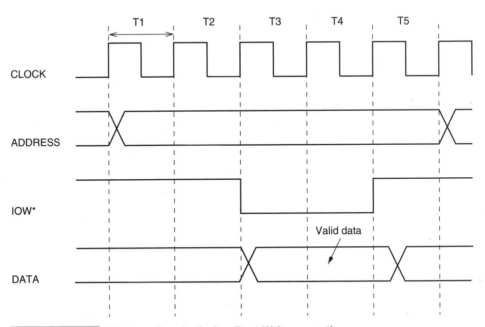

FIGURE 7.21 ISA bus signals during Port Write operation.

mice, keyboard, stylus, game peripherals, phone, audio, etc. It was designed keeping in mind such low- to medium-speed applications.

In this section we are interested in connecting the AVR family of controllers to the USB so as to get a fast interface to the PC. The applications could be to connect a data acquisition system to the PC using the USB, with the AVR as the embedded controller inside the data acquisition system, or to connect a special camera system to the PC through the USB. Figure 7.22 illustrates the possible application space.

The USB is a very complex interface and I do not intend to cover it in detail. Actually, it is complex enough to be the subject of another book. In this section, we just want to facilitate awareness of this interface and identify some other chips that can provide conectivity to the AVR series of processors.

It would be ideal if the AVR processors had a built-in USB port, but that is not the case with the present generation of AVR processors. The USB interface consists of a single signal in a differential format, as illustrated in Figure 7.23. Together with the data signal, the USB cable also carries power supply for use by the peripheral device.

A USB system consists of three components:

1. USB interconnect: This includes the way the devices are connected to each other and to the host and the way these devices share the channel.
2. USB Host: There is only one host in a USB system, which is the host computer.

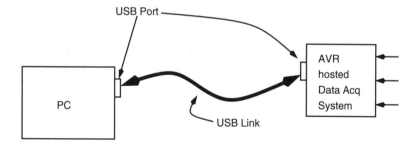

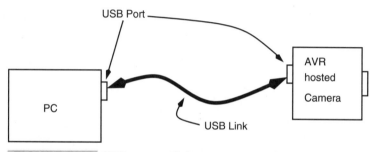

FIGURE 7.22 USB connectivity.

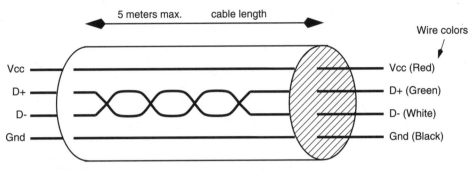

FIGURE 7.23 USB cable.

3. USB devices: These are the actual function devices (like the mouse) as well as hubs, which are like extension points for function devices.

Figure 7.24 illustrates the USB bus topology, i.e, the way all the components connect to the PC that is the host. Figure 7.25 shows a hub and how it can connect to the upstream port (the host or another hub) on one end and to actual devices or other downstream hubs on the other end.

The communication on a USB occurs between point-to-point segments over the two wires. The USB signaling is of 2 types: full-speed signaling at bit rate of 12 Mbit/s and low-speed signaling at 1.5 Mbit/s. Both modes can coexist in the same USB system by mode switching in a device transparent mode.

The clock for the system is transmitted encoded with the data stream using NRZI scheme. Each data packet has a SYNC field to allow receivers to synchronize their clocks.

There are many USB-capable processors in the market. Although Intel was the first one to design the USB microprocessors, recently they have discontinued these products and sold the technology to Cypress. Cypress is now a market leader with their CY7C63 series of low-speed USB microcontrollers and CY7C64 series of full-speed microcontrollers as well as the acquisition of Anchor Chips company, which makes 8051 core and USB peripheral function chips. However, the Cypress and the Anchor chips are complete microprocessors and not really suitable to interface with AVR chips, though not impossible.

National Semiconductors offers an interesting USB interface chip USBN9602, with a parallel interface as well a MICROWIRE interface specifically to connect to general-purpose processors, and from the data sheets, seems a very good candidate to connect to the AVR (Figures 7.26 and 7.27.)

Another USB controller that goes with a local processor is the NET2888 USB chip from Netchip (www.netchip.com) and can be used together with an AVR processor.

7.10   IrDA Data Link

IrDA stands for Infrared Data Association, which has a charter to create a standard for low-cost Infrared data interconnection. This standard is called IrDA data link and allows a walk-up, point-to-point method of data transfer.

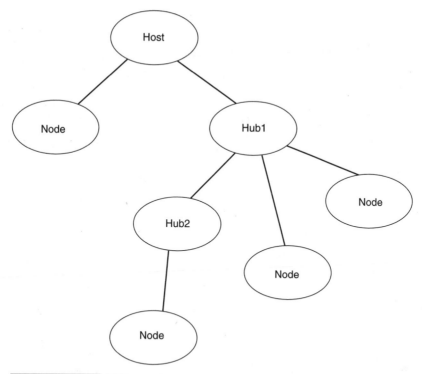

FIGURE 7.24 USB topology.

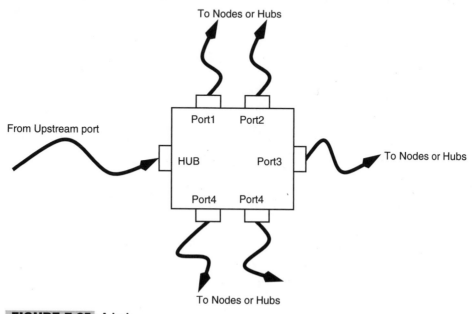

FIGURE 7.25 A hub.

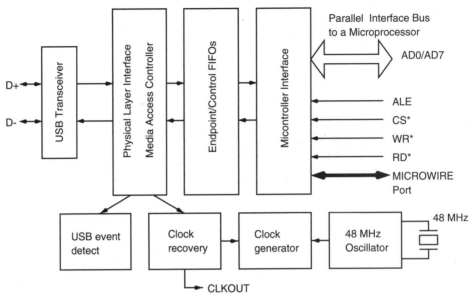

FIGURE 7.26 USBN9602 block diagram.

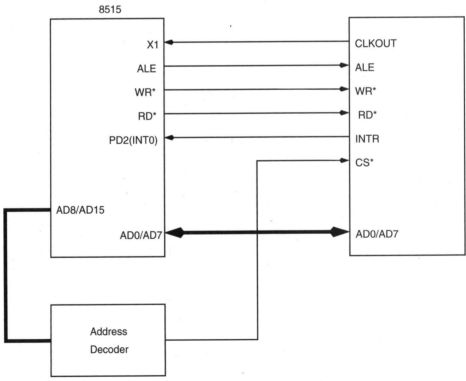

FIGURE 7.27 USBN9602 Interface to AT90S8515 AVR Controller.

The IrDA link works by sending an encoded data stream serially using an infrared LED. For receiving, it has an infrared receiver diode which feeds a decoder and a serial-to-parallel converter. Figure 7.28 illustrates the physical part of the IrDA data link.

IrDA allows data transfer speeds between 2.4 Kbits/s to 4 Mbits/s. Link speed at startup is always at 9600 bits/s and then can be negotiated to a higher or lower mutually acceptable speed by the transmitter and receiver. Up to the data transfer speeds of 115 Kbits/s, the IrDA encodes the incoming serial data (from the parallel-to-serial converter) using a Return to Zero Inverted (RZI) scheme. This is the version 1.0 mode of communication also called SIR (Serial InfraRead).

Above speeds of 115 Kbits/s and up to 4 Mbits/s, it is Version 1.1 and is called FIR (Fast InfraRed). FIR specifies a total of three data transfer speeds: 0.576 Mbits/s, 1.152 Mbits/s, and 4 Mbits/s. Data transfer speeds of 0.576 Mbits/s and 1.152 Mbits/s use an HDLC type of encoding scheme. At 4 Mbits/s, the encoding is done using a Pulse Positioning Modulation (PPM) scheme. However, like SIR, the initial startup operating speed is 9600 bits/s and in SIR mode. Later, on mutual agreement, the devices could move to higher speeds.

The RZI encoding scheme sends a short optical pulse for every logic "0" for a short duration and no pulse for a logic "1." The duration of the optical pulse is 3/16-bit time or 1.6 μs (which is 3/16 of bit time at 115 Kbits/s) (Figure 7.29).

To equip an AVR device with IrDA capability, one needs an IrDA encoder, IR driver, receiver, and IR LED and diode, as illustrated in Figure 7.30 connected to the integrated serial port available in most AVR processors.

The MAX3100 universal asynchronous receiver transmitter (UART) is specifically optimized for small microcontroller-based systems. It is suitable for connecting an AVR processor to MAX3100 using an SPI/Microwire interface. The MAX3100 has an IrDA SIR Timing Compatable, and it only needs an external LED driver and receiver to make a complete IrDA link, as illustrated in Figure 7.31.

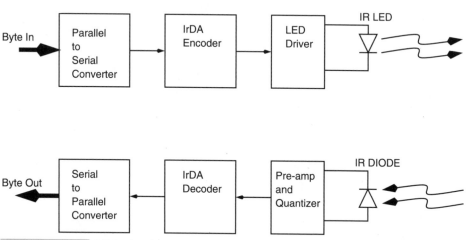

FIGURE 7.28 IrDA physical layer block diagram.

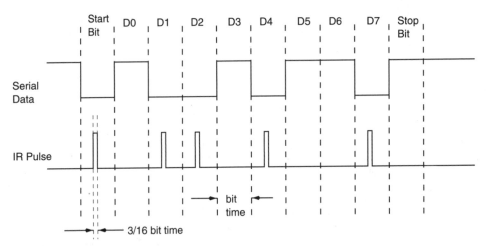

FIGURE 7.29 RZI data encoding scheme employed by IrDA data link.

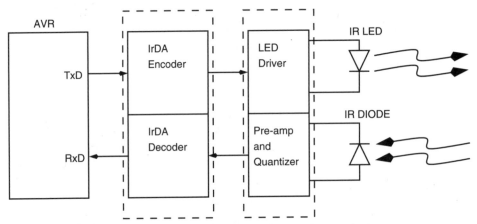

FIGURE 7.30 An AVR processor with an IrDA data link.

7.11 CAN (Controller Area Network) Bus

The CAN bus is a serial communication link used in automobiles as a means of communicating between the various controllers used inside cars and other vehicles. As we mentioned, a modern car has over 50 processors. The CAN bus allows all these processors to talk to each other on a single link, thus reducing cable length, which would otherwise get out of control. Without the CAN bus, the cable length in a modern car could be as much as a few thousand meters.

The CAN bus was designed by a German company, Bosch, originally for automotive use. However, it is being also used as a general-purpose industrial bus (Figure 7.32).

The CAN bus was designed with the following needs in mind:

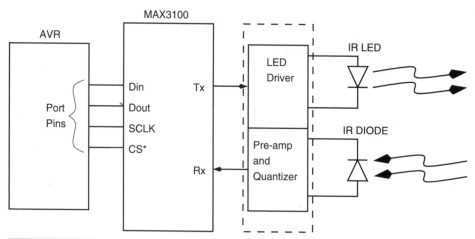

FIGURE 7.31 An AVR processor interface to MAX3100.

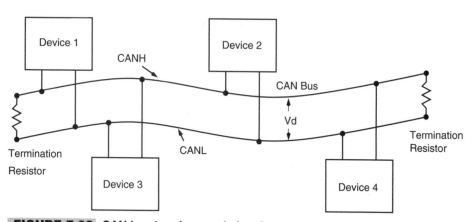

FIGURE 7.32 CAN bus topology and signals.

1. Support low- to high-speed data transfer rates.
2. Provide error-free data transfer.
3. Allow variable data volume transfers.
4. Offers ease of maintenance and low cost.

Philips has introduced a stand-alone CAN controller chip SJA1000, which can in principle be interfaced to an AVR device.

AVR SYSTEM DEVELOPMENT TOOLS

Till now we have considered the internals of the AVR processor in some detail, as well as looked at the ways the AVR processor can be used to connect to the external world.

However, the big question is how to get the AVR processor to actually do something that you want. How do you develop the application and get the system working? What are the tools available to complete the design on time and within budget? This chapter will answer some of these questions.

What are the steps in creating a successful system? We have briefly looked at this question in a previous chapter. We need to define the design requirements to begin with. Then we need to identify the hardware that will satisfy these needs. Since we are dealing with the AVR controllers, we assume that you have zeroed in on a particular AVR controller out of the many AVR controllers listed in Table 3.11.

Once you have chosen the controller, you need to understand how to write the code for the AVR controller, how is it tested, and how it is eventually loaded in the system.

8.1   Code Assembler

What constitutes a program for the AVR controller? Well, the contents of the program memory is the program for the AVR controller. Each memory location in the program memory map of the AVR controller is a 16-bit word. This 16-bit word constitutes the op-code and the (optional) operand. The controller reads the program memory and interprets the binary word. For human convenience, rather than handle the op-codes, you can use the mnemonic representation of the op-codes (e.g., CLI, Clear Global Interrupt flag, is a mnemonic representation of the op-code: $94F0). However, the controller understands the op-code and not the mnemonic, and so you need a translation program that takes the mnemonic codes and translates these codes into op-codes. The program that does this job is called an assembler.

An assembler takes a text file called the source file, with the mnemonic representation of the program (simply called the source code) and converts it into another file with the machine op-codes (simply called the machine code or object code).

8.1.1 AVR FAMILY ASSEMBLER

One of the assemblers available for AVR controllers is from Atmel. It is called avrasm and can be downloaded from the atmel site. The assembler covers the entire range of AVR controllers (Figure 8.1).

The assembler takes an assembler source code file and translates it into an object code file. The object code can be used as an input to a simulator or an emulator. The assembler also generates a file containing the code that can be programmed into the chip by a suitable programmer. The avrasm is a very simple assembler and takes only a single input assembler code file. It is not possible to link other object code files using the avrasm.

The avrasm can also generate an EEPROM file if EEPROM data has been allocated in the assembler source file.

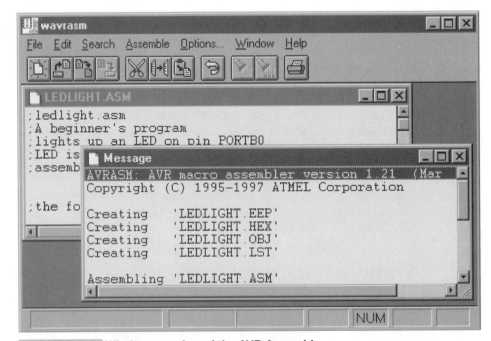

FIGURE 8.1 Windows version of the AVR Assembler.

The avrasm has the following syntax.

```
AVRASM: AVR macro assembler version 1.21 (Mar 5 1998 01:21:00)
Copyright (C) 1995-1997 ATMEL Corporation
usage: avrasm [-m   -i   -g] [-w] <infile> <lisfile> <romfile>
        -m Output Motorola S-record format
        -i Output Intel HEX format (default)
        -g Output Generic format
        -w Relative jumps are allowed to wrap for
          program ROM up to 4k words in size
```

The avrasm runs as an MS-DOS command line version. There is also a windows version available for Windows 3.11, Windows 95, or Windows NT. The windows version of the avrasm is wavrasm, and this has online help. However, the wavrasm cannot assemble very large source files.

8.1.2 IAR ASSEMBLER

IAR assembler is a shareware product from IAR (www.iar.com). This is a very high quality assembler with sophisticated features. The IAR assembler can be downloaded either from the atmel Web site or the IAR site (ftp://ftp.iar.se/pub/mirror/aa90.exe).

IAR assembler allows assembling of multiple files, as well as linking object code. The IAR assembler is fast and has a C preprocessor front end.

8.2   Code Simulator

Now once you have the object file, you can transfer the op-codes into the AVR controller, and if you have written a program exactly according to the system design, it may work correctly. However, there are good chances that your program does not work as expected. In such a case, you will need to find out where the errors are and make suitable changes or additions to the source file, assemble the source file again, and load the machine code into the controller. This may be a time-consuming, iterative process.

Rather than transferring the op-codes into the controller and debugging the code, it is possible to simulate the working of the AVR controller on the development PC itself without downloading the machine code into the controller. A program that simulates or mimics the AVR controller is called a simulator. A simulator can execute the program code, one code at a time, displaying the result on the screen (contents of registers, ports, status, etc.), and this can help ensure that the program works as expected.

8.2.1 AVR SIMULATOR

The AVR simulator can be downloaded from the Atmel Web site. The AVR simulator executes object code files generated for the AVR controllers. It also supports simulation of various I/O functions.

The simulator can be controlled through a command line as well as through menus. The AVR simulator runs under Windows 3.11, Windows 95/98 or Windows NT. The AVR simulator is very easy to use. Figure 8.2 illustrates the simulator window.

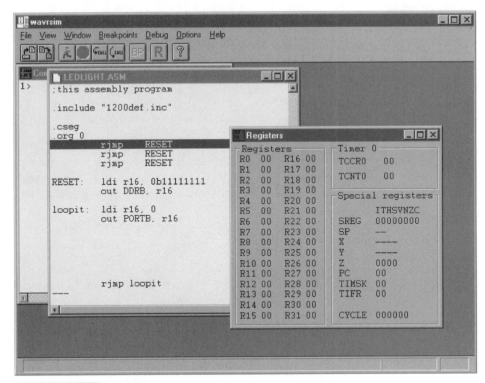

FIGURE 8.2 The AVR Simulator.

8.2.2 AVR STUDIO

The AVR Studio is a development tool for AVR controllers. It allows control of execution of programs in the AT90S In-Circuit Emulator or on the built-in AVR instruction set simulator (Figure 8.3).

The AVR Studio supports source-level execution or AVR Assembler programs as well as C programs compiled with IAR systems' ICCA90 C compiler. The AVR Studio runs under Windows95/98 and Windows NT.

At the time of starting the AVR Studio, if the AVR In-Circuit Emulator is connected and powered on, the AVR Studio detects it and enables execution of programs on the emulator. Otherwise, it invokes the built-in AVR instruction set simulator for source-level simulation.

8.3   Evaluation Boards

Atmel has introduced a few evaluation boards at very competitive prices. It is strongly recommended to acquire one of these boards, depending upon your application and need.

Using an evaluation board you can quickly test your programs before committing a final design board. All of these evaluation boards have a communication link to the PC for downloading your program to the target processor. Since all the AVR controllers can be programmed without the need for a special programmer, the downloaded program is programmed into the target controller program memory by the evaluation board and then exe-

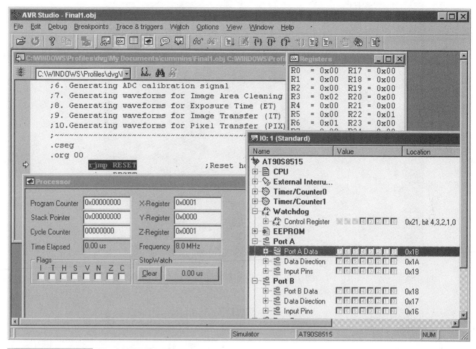

FIGURE 8.3 The AVR Studio.

cuted. The evaluation boards can be used not only for testing your programs, but also as a simple device programmer.

8.3.1 ATMEL AVR MCU00100 DEVELOPMENT BOARD

The AVR MCU00100 development board was the first evaluation board introduced by Atmel. It sells for $49. The evaluation board connects to the PC through a serial port. The board has a resident controller (an AT89C2051 chip) that communicates with the PC, accepts downloaded programs from the PC, and programs the target AVR controller.

The board has a set of LEDs and switches that can be connected to the AVR controller ports. This evaluation board can be used with most 20-pin and 40-pin DIP package AVR controllers, as illustrated in Figure 8.4. The socket for a 20-pin IC is inside the 40-pin socket.

This board has a software package that includes, among other programs the avrprog.exe file that is a Windows 95/98/NT programming interface to the MCU00100 evaluation board (Figures 8.5–8.7).

8.3.2 STK200 BOARD

The STK200 Starter Kit is a complete evaluation platform for developing an AVR design. The STK200 Starter Kit is the ideal method to get acquainted with the AT90S AVR microcontroller family, including the new analog parts. The STK200 Starter Kit from Atmel supports the following Atmel AT90S AVR devices:

1. AT90S1200
2. AT90S2313

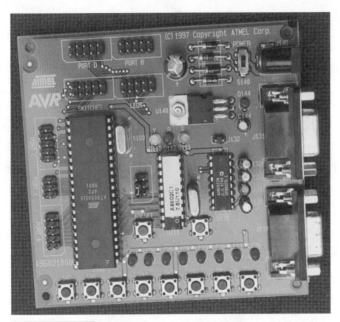

FIGURE 8.4 Photograph of the MCU00100 evaluation board.

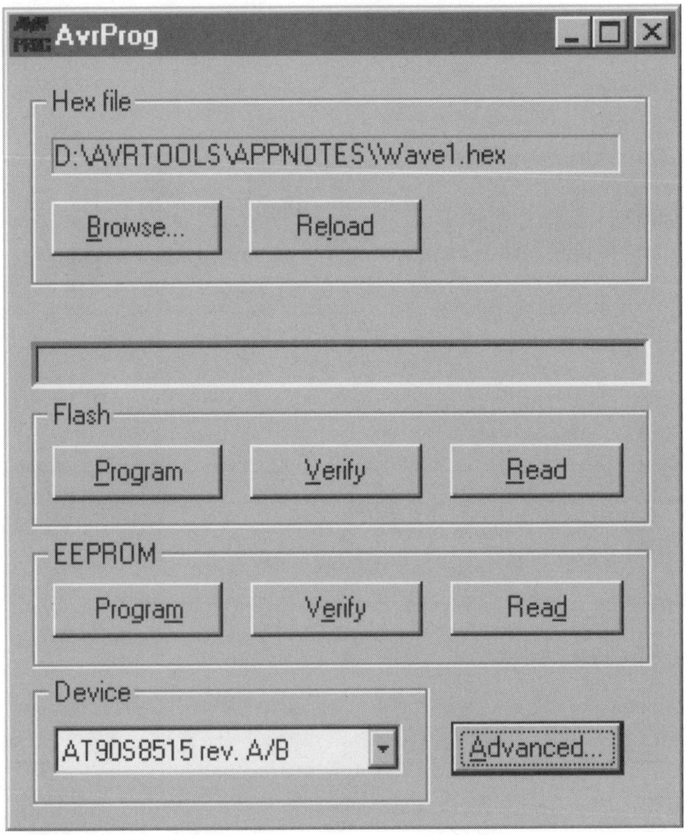

FIGURE 8.5 AVRPROG primary window.

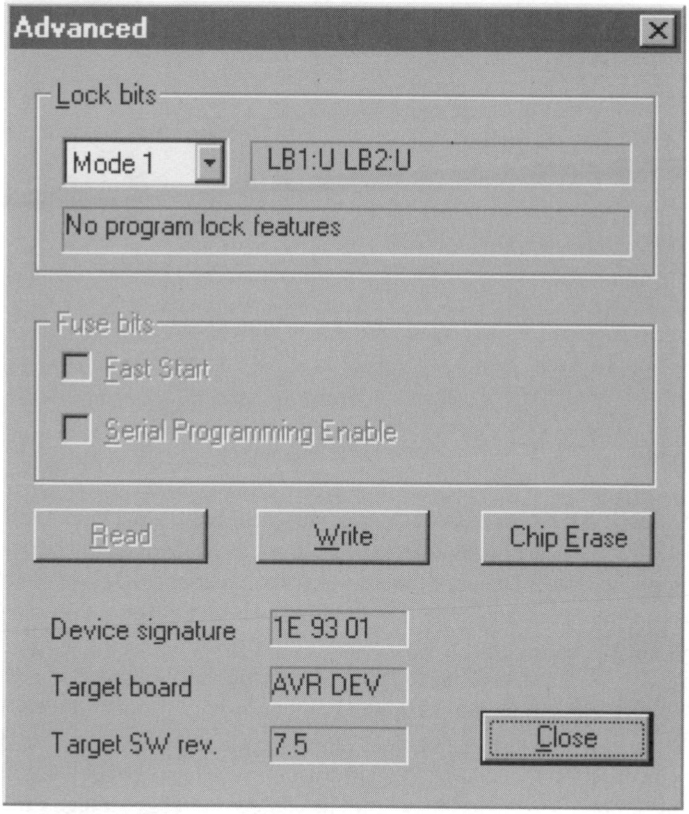

FIGURE 8.6 AVRPROG advanced window.

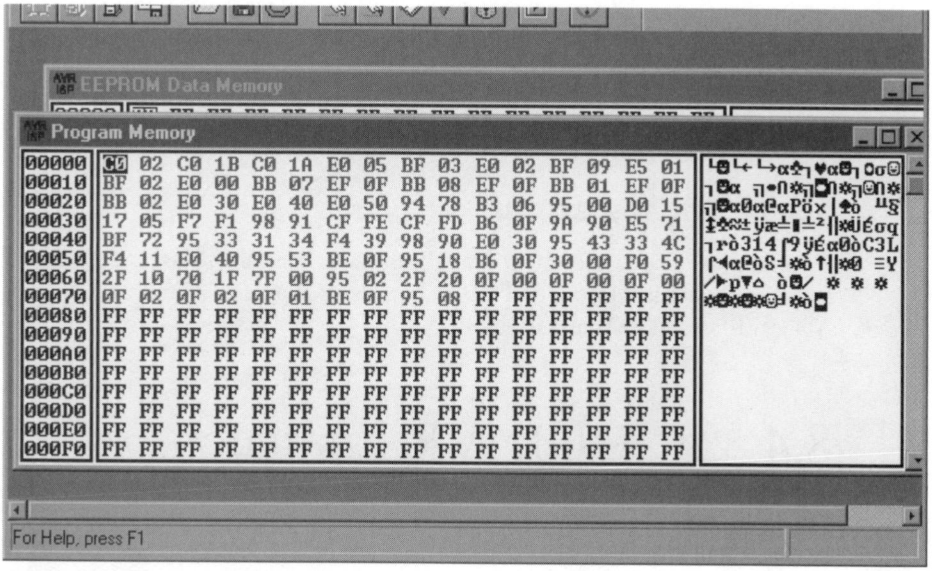

FIGURE 8.7 AVR ISP software.

3. AT90S2323
4. AT90LS2323
5. AT90S2343
6. AT90LS2343
7. AT90S2333
8. AT90LS2333
9. AT90S4414
10. AT90S4433
11. AT90LS4433
12. AT90S8515
13. AT90S8535
14. AT90LS8535

The STK200 development board is complete with push-button switches and LEDs. All of the microcontroller ports are accessible through 0.1-inch headers on the board. The regulated power supply accepts a DC or AC input voltage source, and 3.3-V or 5-V operation can be selected by jumper settings. The on-board brownout detector circuit is user selectable to 2.9 V or 4.5 V.

An external LCD can be interfaced to the AVR microcontroller via a standard 14-pin header on the board. A potentiometer is supplied for contrast adjustment. The STK200 has support for external SRAM with sockets for Address latch and SRAM device. The internal 10-bit A/D converter on the analog AVR microcontroller is accessible through a separate header connector on the edge of the board. The analog reference voltage can be adjusted with an on-board potentiometer or optionally with an external reference voltage.

The STK200 board includes an In-System Programming (ISP) cable that connects the STK200 board to the PC through the PC parallel port. The ISP cable can also be used to program all classic Atmel AVR microcontrollers in the actual target application via a 10-pin header. Figure 8.8 is a photograph of the STK200 evaluation board.

Support site from Atmel: AVR 8-Bit RISC Support Tools http://www.atmel.com/atmel/products/prod202.htm

STK200 Starter Kit Manual (40 pages, updated 2/99): http://www.atmel.com/atmel/acrobat/doc1107.pdf

8.3.3 STK 300 BOARD

If you are planning to use any of the AVR mega devices, then you need the STK300 board. It is very similar to the STK200 but only supports ATmega603, ATmega603L, ATmega103m, and ATmega103L.

8.4   ICE200 AVR Emulator

Atmel has recently introduced a low-cost In-Circuit Emulator, the ICE200, priced at $200. An In-Circuit Emulator has a pod that is connected to the target controller socket during

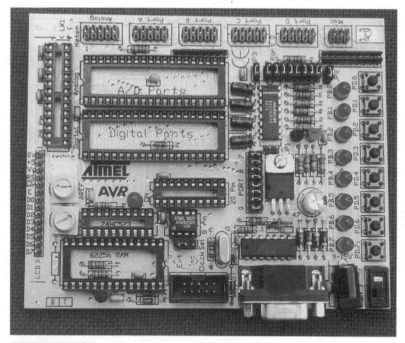

FIGURE 8.8 Photograph of the STK200 evaluation board.

development. The pod connects to the emulator, which has a special "Super" AVR chip. On the other end, the emulator connects to the PC and the user can single-step the program in the target, monitor I/O activity, register contents, set break points, etc. The ICE200 AVR Emulator is controlled through the AVR Studio.

8.5   The Device Programmer

Evaluation boards are great for developing your application as well as programming the AVR controller chip with the code. However, all the evaluation boards program the AVR chips in the so-called low-voltage serial programming mode. The low-voltage serial programming mode cannot program certain fuse bits like the SPIEN and RCEN fuse bits. To program or change these bits, the AVR chip must be programmed in the parallel programming mode (for AVR chips with more than 8 pins) or the high-voltage serial programming mode (for 8-pin AVR chips).

A dedicated device programmer with parallel programming (or high-voltage serial programming) features is very useful to satisfy these needs. The Atmel Web page has a list of many programmer vendors: (http://www.atmel.com/atmel/products/prod205.htm). One of the programmer vendors, MITE (http://www.mite.cz/), offers one such programmer that has had good user response.

8.6 AVR System Design with Components off the Shelf (COTS)

8.6.1 The SimmStick magic

SimmStick™ is a 30-pin simm-socket-based Single Board Computer designed by Antti Lukats from Estonia, originally for the PIC16Cxx chips. The simm socket used is the common 30-pin one used for many years on PC motherboards.

SimStick is now available for Atmel AVR processors for both 20- and 40-pin DIP packaging. The SimmStick system consists of a motherboard (the backplane). The motherboard (such as DT003) has space for up to four daughtercards such as the DT103/104 controller card, additional daughtercards such as DT201 (one-inch prototype board), DT202 (two-inch prototype board), DT203 (LEDs, switches, ULN2803 drivers board) or the DT205 (relay board for four relays).

SimmStick Bus Pinout

PIN #	NAME	DESCRIPTION
1	A1	Special IO
2	A2	Special IO
3	A3	Special IO or Negative Supply
4	PWR	Unregulated DC in or +12V or VPP
5	CI	Clock Input or OSC1
6	CO	Clock Output or OSC2
7	VDD	+5 V In or Out
8	RES	Reset In or Out
9	GND	Digital Ground
10	SCL	I2C Clock or I/O
11	SDA	I2C Data or I/O
12	SI	Serial In or I/O
13	SO	Serial Out or I/O
14	IO	General-purpose I/O
15	D0	General I/O
16	D1	General I/O
17	D2	General I/O
18	D3	General I/O
19	D4	General I/O
20	D5	General I/O
21	D6	General I/O
22	D7	General I/O

23	D8	General I/O
24	D9	General I/O
25	D10	General I/O
26	D11	General I/O
27	D12	General I/O
28	D13	General I/O
29	D14	General I/O
30	D15	General I/O

The advantage of adopting the SimmStick approach is the great speed of system development. It is claimed (and rightly so) that a simple working AT90S1200 application can be wired on the DT104 board in under five minutes.

Figure 8.9 illustrates the DT104 circuit schematic, Figure 8.10 illustrates the DT104 PCB component overlay, and Figure 8.11 illustrates a populated DT104 photograph.

More details about the SimmStick boards are available at www.dontronics.com.

8.7   Code Development with a High-Level Language

For larger AVR processors, using a high-level language for developing your application would be very useful and convenient. Currently, high-level language compilers for C as well as Basic are available. For the C language, IAR's C compiler (www.iar.com) is reported to be the best and also the most expensive. Many inexpensive or free compilers are available, and some of these are listed below.

8.7.1 C-AVR: A C COMPILER FOR AVR

C-AVR is a low-cost C compiler that I have used to develop code for the dual-channel voltmeter project in this book. The features of this compiler are:

1. Supports all members of the AVR family which have stack. e.g. 2313, 2323, 4414, 8515, 8535, etc., and ATmega603/103. (Also available is a low-cost version, C-AVR-N, which has the same features but does not support mega devices.)
2. Uses (and includes) Atmel's AVRASM assembler, so output of the compiler can be debugged with Atmel's AVR Studio debugger.
3. Supported data types: char, unsigned char, int, unsigned int, long int, unsigned long int, float.
4. Runs off Windows 95/98.
5. Includes Visual Code Generator (VCG). Using the VCG, on-chip peripheral functions (such as a serial port) can be visually programmed. E.g., you may just enter the crystal frequency and the desired baud rate; the VCG generates C language statements such as "UBRR = 0x18" and "UCR = 0x20," etc.

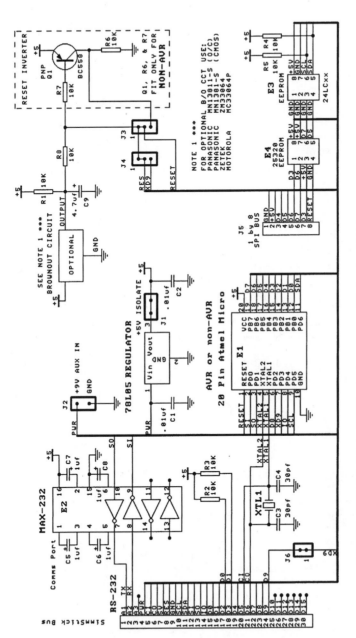

FIGURE 8.9 DT104 schematic.

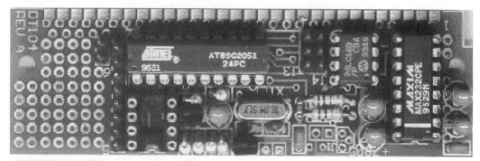

FIGURE 8.10 DT104 component overlay.

FIGURE 8.11 Fully populated DT104 board.

6. Function library includes many useful functions, including printf, scanf, etc. You can add functions to the library or create your own libraries.

7. Package includes an Integrated Development Environment (IDE), C Compiler, assembler, linker, library manager, and some example programs. Most example programs are written for and tested with the Atmel AVR Starter Kit (STK-100)

8. Software upgrades are free of cost for an unlimited period and can be downloaded from the Web site.

A demo version of C-AVR can be downloaded from the http://www.spjsystems.com Web site.

8.7.2 DDS MICRO-C DEVELOPERS KIT FOR THE AVR

DDS MICRO-C developers kit is a complete PC-based cross development system that includes everything to develop C and assembler software for the AVR. The kit includes MICRO-C compiler, optimizer, XASM cross assembler, and related utilities for the AVR. Integrated Development Environment that includes an editor, compiler, code download, and a debugger. Complete documentation is available on a disk. The kit sells for $99. More information is available at this Web site: http://www.dunfield.com/dks.htm.

8.7.3   BASICX: A BASIC INTERPRETER FOR THE AVR

The BasicX-1 microcontroller is a one-chip computer that runs programs written in Basic. The BasicX-1 has 32 fully programmable I/O pins that can be directly interfaced to a multitude of devices. Examples are LEDs, SPI devices, potentiometers, card-swipe readers, switches, buttons, LCD displays, and keypads.

And just by adding a few extra components, you can interface non-logic-level devices as well, such as relays, electric motors, blowers, air solenoids, and stepper motors.

The BasicX-2 is an AT90S8515 chip. A BasicX-1 chip requires only an external crystal, 2 capacitors, and source of 2.7 to 6.0 VDC power. After downloading your program to a BasicX-1 system, the compiled program is extracted at startup by a built-in high-speed interpreter. The interpreter converts your code into machine language executed by the BasicX-1's AT90S8515 controller. The BasicX-1 can execute a program at a rate of about 65,000 lines of code per second.

8.7.4   BASCOM-AVR: A BASIC COMPILER FOR THE AVR

BASCOM-AVR is a Windows BASIC COMPILER for the AVR family. It is designed to run on Windows 95/Windows 98/Windows NT platforms.

BASCOM-AVR is a structured BASIC with labels, IF-THEN-ELSE-END IF, DO-LOOP, WHILE-WEND, SELECT-CASE. BASCOM-AVR runs on all AVR processors with internal SRAM. Statements are highly compatible with Microsoft VB/QB. The compiler offers special commands for LCD-displays, I2C chips. The compiler has an integrated terminal emulator with download option, an editor with statement highlighting, and context-sensitive help. There are plans for an integrated simulator for testing purposes.

The compiler sells for $69. More information is available at this Web site: http://www.mcselec.com/bascom-avr.htm.

8.7.5   JAVRBasic: JACK'S AVR BASIC COMPILER

Jack Tidwell has developed a Basic compiler for the AVR, and it is available for free. The compiler is under development, and the latest version can be downloaded from the author's Web site: (http://www3.igalaxy.net/ jackt/ and http://www.javrbasics.com).

JAVRbasic is a small Basic compiler for any Atmel AVR processor with internal SRAM. The author has kept the JAVRBasic syntax close to that of VB, and it is possible to mix assembler commands inside the basic program.

9

PROTOTYPING TECHNIQUES

If you are like most people, eventually you would like to build your system. The usual steps in completely designing a system is to put your ideas on a piece of paper, estimate the required hardware, draw some flow chart of the software, simulate the design on any available tool, and then finalize the design. Eventually it will come to a stage where you must build it and test it. For those who earn a living from electronic design, you must also think of production plans for your design.

At this point, we will consider various options that are popularly used to prototype electronic circuits and systems. I have put together a list of tools that I have found useful in prototyping my projects. I also offer some suggestions on some soldering tools for various applications.

9.1 Why Prototype?

An important questions could be raised at this stage. Why prototype the design at all? What does prototyping the design gain you? After all, you may have to make the PCB for the design anyway? Why not go directly for the final PCB?

Prototyping allows you to test your design and evaluate its performance at a very little cost and rapidly. If you have gone through the process of making a PCB, you will realize that this time-consuming process of capture, board layout, and fabrication will delay the crucial task of verifying your design. Besides, the design itself may be in a preliminary stage and you may want to add and modify the design and see the performance on a proto board. So prototyping your design is a good idea. For hobbyists, the prototype could be the final design anyway.

The importance of design verification while still on the drawing board (or the design board—whatever) cannot be overemphasized. It is a good idea to thoroughly check the design while it is still on the paper. Another way to check your design is to simulate the performance of the complete or part of the circuit design on any simulation tool that is available, say a code simulator. However, the prototyping stage of the design will give you one last opportunity to check that the performance is according to specifications and is bulletproof, and hopefully you will be saved the embarrassment of a failed and returned product from the field or from the customer.

There is a possible downside to prototyping as well. Many times there is an urge to rush your circuit and see it working as a prototype without giving it much thought while it is in the design stage. It is advisable to control this urge and to prototype the idea when you are satisfied that the design has been well-thought out and suitably simulated. A badly designed circuit that has been rushed to the proto stage could end up in an endless and vicious loop of modify-prototype-test. This must be avoided at all costs.

9.2   OK, So You Want to Prototype

There are many ways of prototyping your circuit. For very simple circuits, it is popular to use the protoboard. The protoboard is made of plastic (or some such material) with holes at a 0.1-inch pitch in the form of a regular matrix. The holes in a given row are internally connected. The holes are of a certain diameter that accepts most component leads such as small diodes, resistors, transistors, and ICs in DIP packages, etc. On either side of the protoboard, there are two columns of connected holes that are usually used as a power bus to connect the protoboard to a DC voltage source for energizing the circuit. Figure 9.1 is a photograph of a popular protoboard.

To use the protoboard, one simply pushes the components in the holes and makes required connections with suitable single-strand hookup wire. The protoboard is a simple and convenient way to test your circuit. And it does not require you to dirty your hands with the soldering rod. However, the protoboard is not suitable for high-speed applications. The inductive coupling between the rows of connections may produce unwanted and undesirable crosstalk.

Another simple method is to use the commonly available general-purpose board made of glass epoxy, the material that the PCB is usually made of. This board, usually referred to as a "sea of holes," is available in many formats form many suppliers. Figure 9.2 is a photograph of a general-purpose PCB. This general-purpose circuit board has holes arranged on a 0.1-inch pitch with solder contacts on one side. These boards are also available with many connector pinouts as well, to connect to say a DB-25 connector. Prototyping boards for the PC interface connectors are also available.

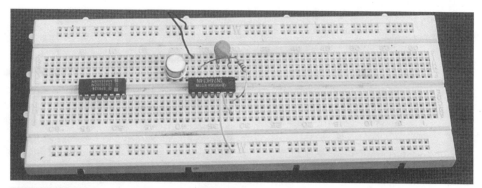

FIGURE 9-1 Photograph of a Protoboard.

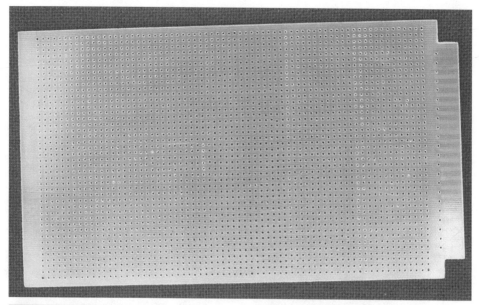

FIGURE 9-2 Photograph of a general purpose printed circuit board.

Another method for prototyping your circuit with a minimum of effort is to buy an evaluation board for the processor that you want to use in your design. Usually, these evaluation boards have some general-purpose prototyping area with a matrix of holes. The advantage of using the evaluation board is that you can use the resources on the board to get your design up and running. This usually means building your specific circuit portion of the design. The processor part of the design is used from the evaluation board. Another advantage is that these boards usually come with a collection of useful routines that you can plop into your design (I use the word "design" to refer to the composite hardware software portions). Usually, you also get an integrated debugger that eases your design test. This is not completely true for the AVR family of controllers because the program memory is located inside the controller chip. However, you do get the advantage of a working controller core circuit. The downside of this approach is the cost of the evaluation board.

9.3   Tools of the Trade

Whether you want to use the evaluation board to prototype your design or you want to use the general-purpose PCB, you will need a set of suitable tools to do a good job. It is very important that even the prototype is fabricated with the same care as the final product, if not more. You don't want some silly, sloppy solder job to be a cause of a nonfunctional design. Since you are testing a new design, you want the influence of other variables to be minimal (like a badly fabricated prototype with shorted interconnections). To do that you need to employ suitable tools to ease the construction. Figure 9.3 is a photograph of some useful tools, including the wire stripper supplied by RS components.

These tools are:

1. Solder iron, 35 watts, with a fine solder tip. A soldering station is highly recommended but is not mandatory. The soldering station offers isolated supply to the solder iron heater, thus reducing the leakage currents from the tip of the solder iron. Such a configuration is useful for use with CMOS components. An ordinary soldering rod can also be used while soldering susceptible CMOS components by temporarily disconnecting the main supply to the solder iron (when it has reached its operational temperature) just at the time of actual soldering and then connecting the supply back again.
2. Fine tweezers for bending component leads.
3. Nipper to cut the component leads. This is probably a fancy name for the regular lead cutter. A nipper has sharp edges that make a neat cut.
4. Wire stripper (more on this in the text). A wire stripper is very handy in stripping a precise amount of wire insulation.
5. Nose plier. Generally useful for tightening screws, etc.

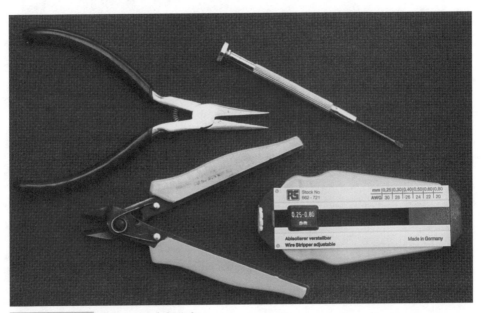

FIGURE 9-3   Some useful tools.

6. Screwdriver set.
7. M3 nuts and bolts. For fastening brackets onto the PCB as well as to support the PCB.
8. Drill machine (hand operated will do), with an assorted collection of drill bits. Use for drilling holes in the PCB, enclosures, etc.
9. Bench vice with a 3-inch jaw. For holding the PCB steady, filing hardware or PCB, etc.

9.4 Steps for Prototyping

Now you are ready with all your tools and have your design ready. We can begin by looking at a procedure for prototyping any given circuit. It is important to have all the required components at hand.

I usually proceed with the following steps when I am putting together a circuit for the first time:

1. Estimate layout for the circuit. It is elementary to keep components that share connections closer to each other. This will keep the connections short. How close should the components be placed? This is a tough question. Usually, this estimate is intuitive. As an example, two DIP ICs should be placed .2 inch or more apart, depending on the number of interconnections to other components. If connections allow, even a .1-inch separation could be used.

2. After you make an estimate of the placement of components and connectors, it is time to cut a suitable piece of the general-purpose "sea of holes" PCB. Cutting a piece of PCB a little more than you estimated is not a bad idea. This would take care of any future additions that you may wish to have. Now drill four M3 holes on the corners of the PCB.

3. The four M3-size holes on the PCB corners are populated with 1-inch-long screws such that the PCB is raised about .5 inch above the ground. The other .5 inch of the screw juts above the PCB. With this arrangement, the PCB can be inverted for ease of soldering on the solder side of the board without putting stress on the components on the component side.

4. Usually it is a good idea to start mounting components that have the smallest height profile. Neatly bend the component (resistors, diodes, etc.) leads at right angles and of the right length to slip into the PCB holes. To save space, it is also common to place components like resistors or diodes in a standing position. This requires that the other lead of the component is bent and looped back into an adjacent hole on the PCB.

5. Since this is a prototype, it is all right to use sockets for the ICs. Choice of sockets is crucial, and nothing is gained by going in for cheap alternatives. IC sockets and connectors in general have high failure rates and are best avoided for critical applications. This is a choice that you must make on your own. All other things being equal, a circuit with minimum sockets and connectors is usually more reliable than one which has more sockets, connectors, etc. The downside is the pain in desoldering and replacing any IC.

6. Once you start plopping components onto the PCB and soldering them in place, you have to worry about interconnecting them. The most common method of interconnection is to use ordinary plastic insulation single or multibraid hookup wire. However, we

at IUCAA prefer single, tinned copper wire with Teflon insulation. The advantage of using Teflon insulated wires is that the insulation is very sturdy and can be routed in any manner, between IC pads, on the PCB without having to worry about damage while soldering other components. A Teflon-insulated wire is also more mechanically robust than the plastic counterpart. A major difficulty with Teflon wires is in stripping the Teflon insulation. To strip these Teflon-insulated wires efficiently, a special stripping tool is available from Radio Shack and other vendors. This stripping tool has an adjustable strip diameter and length and can be used from all wire gauges from 10 to 30 SWG (or corresponding AWG).

7. It is useful to put in suitable test points on the prototype to ease probing with a voltmeter/logic-analyzer/oscilloscope. A general rule for distributing the test points is on the input, output, and control signals of the prototype, selected with discretion.

8. Once the fabrication is complete, a good idea is to verify the connections and to ensure that all the connections are in good shape. Many bad, dry solder joints are known to cause unimaginable problems.

9. At this point, you are ready to apply power and to run some test code. With an onboard microcontroller, a simple beginner's test is to look for a proper system clock. This test does not require any code to be run on the microcontroller. Next would be to see if the system reset functions properly. It is important to catalog the system responses to test code and patterns applied to the prototype. A bottoms-up approach would be to run a series of test codes with increasing integration of system functions.

Figure 9.4 and 9.5 are photographs of component and solder sides of a prototype under construction. Note a tantalum capacitor soldered directly on the pins of an IC in 9.5.

FIGURE 9-4 Component site photograph of a prototype under fabrication.

FIGURE 9-5 Solder side photograph of a prototype under fabrication.

AVR PROJECT 1

SMART DICE: A DICE WITH AN ATTITUDE

10.1 At a Glance

Salient features of this chapter are:

An elementary project that you can build in a couple of hours.

Uses an AT90S1200 and sundry components.

Discusses alternative designs.

Shows how to write code in a variety of ways.

Shows how to modify the circuit and code for an 8-pin AT90S2343.

10.2 Introduction

With this chapter, we begin the last section of the book, that of projects. I will illustrate the various features of the Atmel AVR RISC controllers with a series of practical applications that can be used right away or can be used in a larger project.

I have arranged the projects in order of an increasing complexity. It all starts off with a simple project using an AT90S1200.

The project described in this chapter is an electronic dice. It really is a single-chip project using an AT90S1200, seven LEDs, two switches, and a handful of other components to give a modern-look dice. The circuit and the code can be easily modified for an AT90S2343 to reduce board size and components. In the end, I have touched on this issue of modifying the code and the hardware just a little bit to run off an AT90S2343.

10.3 Design Issues: Specifying the Requirement

Before we embark on any project, it is a good idea to write down in detail what we are looking for. What do we expect and what would it take to achieve our objective? Putting things in black and white is a good habit not only if you are a one-person team but more so if you are working in a larger group.

Now, this project is about building an electronic dice. The device should have an input switch, and upon pressing the switch, a number (selected randomly by the controller) between 1 and 6 should be displayed. The dice display could be built in various ways. You could have a 7-segment display that would display the number (between 1 to 6) when you press a key, or you could have 7 ordinary LEDs arranged as illustrated in Figure 10.1, and as is also the case in a traditional dice.

We would also like to have the display blank off momentarily after the switch for when a new number is pressed. This gives feedback to the user that the switch press was recognized by the processor. In the absence of this blanking feature, if the user presses a switch and the next number happens to be the same as the last one, the user may not recognize that.

Other requirements could be that the circuit should consume as little power as possible and should be as compact as can be.

To minimize the power consumption requirement, we could do two things:

1. Choose high-efficiency LEDs that require less current for operation, or you could choose the largest possible series resistor to limit the current through the LED for just sufficient light.
2. Another method to minimize power consumption is to blank off the display if the dice

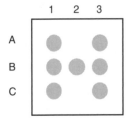

FIGURE 10.1 Output LED arrangement for our dice.

is not used for more than a certain time, say 5 seconds. Thus, if the user does not press the input switch for more than 5 seconds, all the LEDs should be put off. Of course, the circuit is alive and would respond to a switch press even if the display is blanked off. Another level of reduction in power consumption could be to automatically shut down the circuit if it is not used for, say, 1 minute or more. In this mode, the program code executes the sleep mode of operation, which shuts off everything and the processor current consumption is reduced to the lowest possible levels. The only way to revive the circuit is to reset it.

The first method of reducing power consumption is hardware based, while the second one is software controlled. We will look at the second method of reducing power consumption in the section on code development.

At this point we have specified the requirements for our little dice in some detail. We could argue a little more in favor of our design choices. After that, we move over to the next section describing the design in complete detail.

Though we could do with 6 LEDs, for the sake of symmetry, we go in for 7 LEDs. The LEDs are arranged as illustrated in Figure 10.1. Each dot represents an LED. The lighting pattern for the numbers 1 to 6 is as follows:

1: B2
2: B1, B3
3: B1, B2, B3
4: A1, A3, C1, C3
5: A1, A3, B2, C1, C3
6: A1, A3, B1, B3, C1, C3

For each of the numbers between 1 and 6, Figure 10.2 illustrates the LED lighting format graphically. Figure 10.3 illustrates the block diagram for the version of the dice we

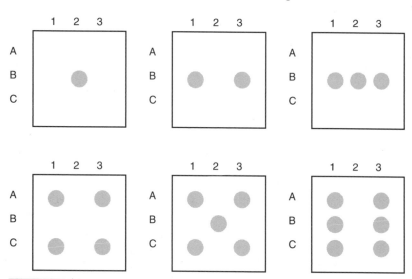

FIGURE 10.2 LEDs light up in this fashion for the numbers 1 to 6.

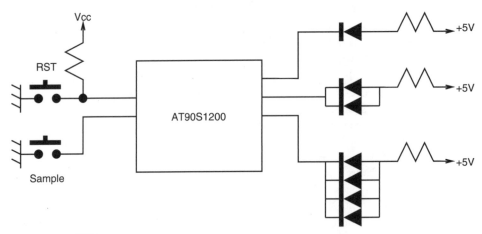

FIGURE 10.3 Block diagram for the electronic dice circuit.

have chosen to implement. In later sections, we will go through other alternative arrangements that may appeal to you, and you could choose to implement code for those designs. For now let us consider the advantages of our current design. The block diagram shows that the 7 LEDs are arranged in blocks of 1, 2, and 4 LEDs. Bunching the LEDs this way allows us to generate all possible display combinations required for the dice. We could also have gone in for 6 LEDs arranged in blocks of 1, 2, and 3 LEDs, but that would not give us the kind of symmetry the 7-LED version provides.

These three groups of LEDs are driven by an output bit of the AT90S1200 processor. The LEDs are arranged such that the LED glows when the output pin from the processor is low (i.e., at logic "0"). This arrangement is usually preferred over the alternative in which the LED glows when the output bit is high (i.e., at logic "1"). For that arrangement, one end of the LED (the anode) is connected to the processor output and the cathode is grounded with a suitable series resistance for limiting the current through the LED. When the LED is lit by setting the output terminal to logic high, the current flows out from the processor output into the ground through the series current-limiting resistor and the LED. This is called sourcing the current.

In the arrangement that we have chosen for connecting the LEDs to the processor output, the cathode is connected to the processor output and the anode is connected to the +ve supply (+5 V in our case) with a series current-limiting resistance.

In this arrangement, to light the LED, the processor output is low and the current from the supply terminal flows into the processor output pin through the LED and the series current-limiting resistor. This is called sinking the current. Typically, a digital gate output (like the processor output pin) is capable of sinking more current than sourcing current.

The block diagram in Figure 10.1 shows two switches marked RST and Sample. The RST switch is used to reset the processor, while the Sample switch is used to interact with the dice. Every time the Sample switch is pressed, the processor gets a random number and displays it on the output LEDs.

The block diagram hides various matters of detail, like the clock frequency for the dice, the way the random number should be generated, etc., which we will discuss in the next section.

10.4   Design Description

Let us consider the details of the final design for the dice, based on the arguments in the last section. Figure 10.4 illustrates the schematic for the electronic dice. The processor we have used is an AT90S1200 in a 20-pin DIP package. In the circuit this is represented by U1.

The reset components for the circuit are a 10-K resistor R4, a 0.1-μF capacitor C4, and a SPST switch J1. As we have seen in a previous chapter, the reset for AVR controllers is active low. So the capacitor is connected from the RST pin of the processor to ground, and the 10-K resistor is connected between the +5-V supply to the RST pin. At power on or at any time the switch J1 is pressed, the capacitor is discharged, and when the switch is released, the capacitor charges up to the supply voltage with a time constant, in the present case, of 1 ms. This holds the RST pin at logic "0" sufficiently long to reset the processor.

The clock for the circuit is derived from a 2-MHz crystal connected between the X1 and X2 pins of the processor. Capacitors C2 and C3 are required to allow the internal oscillator to oscillate. Typical values for C2 and C3 are between 10 and 33 pF. We have used 22 pF for these capacitors.

Switch J2 is the input to the dice, and pressing it produces a new number and is displayed on the 7 LEDs. Normally, when a switch is connected to a processor circuit or a digital circuit in general, it is connected as illustrated in Figure 10.5. The resistor marked R pulls up the switch output to logic "1" and when the switch is not pressed, the logic level at the switch output is "1." This resistor is also called a pull-up resistor. When the switch is depressed, the output is grounded and the logic level changes to "0" till the switch is released.

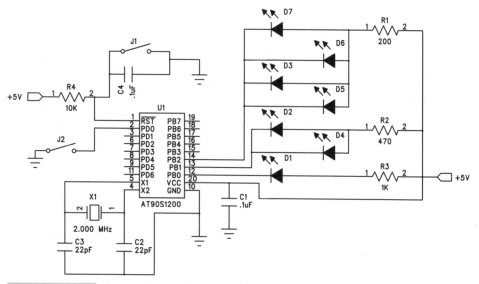

FIGURE 10.4   Schematic for the electronic dice.

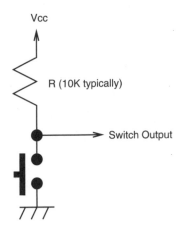

Vcc

R (10K typically)

Switch Output

FIGURE 10.5 A typical switch connection configuration for connecting to processors or a digital circuit.

In the case of AVR processors, the logic circuit on each of the PORTB and PORTD pins can be programmed to have an internal pull-up resistor when the pin is configured as an input pin. This means an external pull resistor for a switch is not required, thus saving component count, board space, and, of course, cost. For our circuit, we have done exactly that—connected the J2 switch to PORTD pin PD0 without any external pull-up resistor.

The 7 output LEDs are connected to the PORTB pins PB0, PB1, and PB2 in groups of 1, 2, and 4 LEDs respectively. A resistor limits the current through each group of LEDs. For the LED connected to the PB0 pin, I have chosen to use a 1-K resistor, R3. This allows about 3.2-mA current through this LED (marked D1). PORTB pin PB1 drives the next group of 2 LEDs, D2 and D4, and R2 with a value of 470 ohms allows a total of 6.4 mA current to be shared by the two LEDs. The third group has 4 LEDs D3, D5, D6, and D7, and a 200-ohm resistance R1 limits the current through these LEDs. To light a group of LEDs, the corresponding port pin should be held at logic "0."

I have chosen to use a common current-limiting resistor for the group of LEDs, though in principle we could use an independent resistor for each LED. However, that would increase the component count.

Capacitor C1 is connected between the Vcc input pin of the processor and ground and is for power supply filtering. The circuit operates at +5-V supply voltage and can be derived from a variety of sources as discussed in a later section.

This completes the description of our design. In the next section, we look at other possible implementations for the dice project.

10.5   Possible Alternatives

Are there any possible alternatives to the circuit we have chosen to implement? Yes, many. Well, at least two.

Figure 10.6 shows a possible alternative. Here we have used a common anode seven-segment display. A resistor for limiting current through each of the seven segments would be required. The processor program would need to change to drive the seven segment display accordingly.

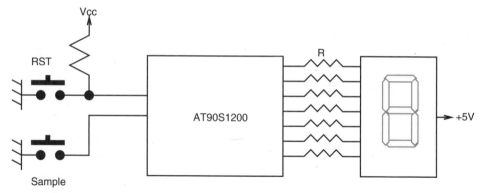

FIGURE 10.6 Block diagram for an alternative electronic dice circuit. R is a current-limiting resistor for each of the segment LEDs of the seven-segment display.

Figure 10.7 shows another possible alternative, not much different from our actual implementation, except that each LED is driven independently by an output pin of the processor. That only puts more demands on the processor resources. For this simple project using an AT90S1200, that may not be a big deal, as the extra required output pins are available, but if you want to port it to a different processor with fewer pins, that may not be possible. Also, this scheme, like the last one, requires extra resistors.

10.6   Code Development

How does one go about developing the whole system to be even as simple as the present one? Does one put together the hardware and then write code for it, or develop code first and then build the hardware? This is a tough question and there is no unique answer to this. It will depend on a particular application.

Generally, it is a good idea to write and test as much code as possible using a simulator or a prototype board. For this application, I used the Atmel's evaluation board, MCU00100 (this has now been superseded by the more advanced STK200 and STK300 boards provided by Atmel) and wrote and tested all of the code. Once the code worked as I wanted, I went ahead and built the circuit on a general-purpose PCB as described in the next section.

The code itself evolved. I present the three versions of code, each with some improvement over the previous version. These three versions of code are put in assembler files named ugly_dice.asm, bad_dice.asm, and good_dice.asm.

The program in ugly_dice.asm was the result of the first attempt at programming the dice. It differs from the code in bad_dice.asm in the way the random number is output on the LEDs. I ended up writing complex code to essentially implement an if-then-else structure. Later I realized that for our simple case, such a complex coding scheme is not required. The code in ugly_dice.asm was then simplified and the resulting code is presented in bad_dice.asm.

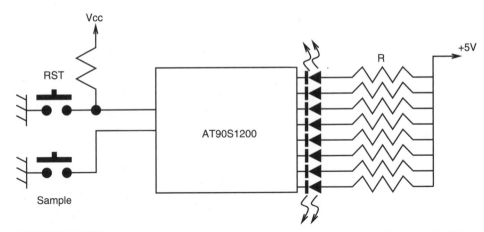

FIGURE 10.7 Block diagram for another alternative electronic dice circuit. R is a current-limiting resistor for each of the LEDs. The seven LEDs are again arranged as in Figure 10.1.

Finally, power-saving features were added to bad_dice.asm, resulting in the final code in good_dice.asm.

Anyway, let us consider the code presented below.

```
;ugly_dice.asm
;
;In the program I just initialize the Timer0 to count up
;using the CLK/1024 as a clocking source. The Timer0 merrily
;counts up, resets to 0 and starts all over again ad infinitum
;PortB is configured as all output and PortD bit 0 (PD0)
;as an input. The program waits for a key press and then
;takes a snap shot of the Timer0 (register TCNT0), and then
;is further processed before put on the display. read on..
;Dhananjay V. Gadre
;19th February 1999
.include "1200def.inc"
.cseg
.org 0
        rjmp    RESET           ;Reset Handle
RESET:  ldi r16, 0b00000101     ;DIV1024 selected for timer0
        out TCCR0, r16          ;timer0 counts up now
        ldi r16, 0b11111111     ;configure PORT B for all outputs
        out DDRB, r16
        cbi DDRD, 0             ;configure pin PD0 as input
        sbi PORTD,0            ;set the PD0 so that it can act as i/p
                                ;with internal pullup
get_t:  rcall sw_input          ;wait for the switch to be pressed
        ldi r18, 0b11111111     ;blank off the previous display
        out PORTB, r18
        rcall delay             ;wait for some time
        rcall delay
        rcall delay
        rcall delay
        rcall delay
        rcall delay
        in r17, TCNT0           ;read the Timer0 value
more:   mov r18, r17            ;copy r17 to r18
```

```
;now check if the number is less than $06
compa:  cpi r18, $06
        brlo enough         ;yes, then nothing more
                            ;prepare to output it
                            ;else
        clc                 ;clear carry for sub
        sbci r18, $06       ;and subtract $06
        rjmp compa          ;do it till the remainder is
                            ;less than $06
enough:    cpi r18, $00     ; OK, now the number in r18 is
                            ; between 0 and 5
;The following piece of code makes this program ugly
;The identification of the number and display can be handled
;more cleverly as in bad_dice.asm
        brne not_0          ;number is not zero
        ldi r18, 0b11111110 ;number is 0, so display 1 on the
        out PORTB, r18      ;LEDs, i.e. light up LED B2
        rjmp get_t
not_0:  cpi r18, $01        ;check if the number in r18 is 1
        brne not_1
        ldi r18, 0b11111101 ;yes, so light up B1 and B3
        out PORTB, r18
        rjmp get_t
not_1:  cpi r18, $02        ;check if it is 2
        brne not_2          ;no it is not
        ldi r18, 0b11111100 ;yes, it is. so light up B1 and B3
        out PORTB, r18
        rjmp get_t
not_2:  cpi r18, $03
        brne not_3
        ldi r18, 0b11111011
        out PORTB, r18
        rjmp get_t
not_3:  cpi r18, $04
        brne not_4
        ldi r18, 0b11111010
        out PORTB, r18
        rjmp get_t
not_4:  ldi r18, 0b11111001
        out PORTB, r18
        rjmp get_t
;Delay subroutine
;uses registers r16 and r18
;values are set arbitrarily
delay:    ldi r16, 0b11111100
loophere:  ldi r18, 0b10001110
decrement: dec r18
        brne decrement
        dec r16
        brne loophere
        ret
;returns when a key on PD0 is pressed and released
;till then it loops waiting for the key press
sw_input: in r16, PIND      ;input PORTD pin value
        andi r16, 0b00000001 ;isolate PD0 state
        brne sw_input       ;if switch is not pressed, loop back
        rcall delay         ;if pressed, then wait some time to
                            ;ward off the switch bounces
pin_0:    in r16, PIND      ;now check if the switch is released
        andi r16, 0b00000001
        breq pin_0
```

```
        rcall delay
        ret                     ;switch is now released..go back
```

The code in bad_dice.asm is the same as in ugly_dice.asm, except for the part that displays the resulting number (between 0 and 5). This code is as follows:

```
;code segment from bad_dice.asm
enough: ldi r16, $01           ;since the number is between 0 and 5
                               ;add '1' to make it between 1 and 6
        add r18, r16
        com r18                ;the output display LEDs are arranged
                               ;as active low, so complement the result-
                               ;ing
                               ;number
        ori r18, 0b11111000    ;set the other unused port pins
                               ;to an inactive state
        out PORTB, r18         ;display it!
        rjmp get_t             ;that it! go get more
```

The program uses free-running Timer0 to get the randomness. However, it is possible to use other methods of random number generation; a prominent one is to use Linear Feedback Shift Register (LFSR) as outlined in a previous chapter. The advantage of the LFSR method is that it only requires a seed to arm the LFSR, and then it will churn out random numbers.

For this simple case, where the Timer0 is not going to be used for any other task and is always available, we could use it full time. The Timer0 is clocked at 1/1024 the clock frequency, and since the state of the timer is not known to the user, reading it gives a certain randomness suitable for our purpose.

The program in good_dice.asm builds upon the code in bad_dice.asm by adding an interrupt subroutine that occurs every time TCNT0 overflows. With a clock input to the counter of clk/1024, this occurs every 128 ms for a 2.000-MHz clock. The initialization code in good_dice.asm initializes three registers as low_timer, med_timer, and high_timer to zero. The interrupt subroutine increments these registers each time the timer overflow occurs.

By looking at the values in med_timer (you can modify the Timer0 ISR so that it powers down after a longer time), the Timer0 ISR determines whether to blank off the display with power-down sleep or not.

The Timer0 interrupt subroutine from good_dice.asm is illustrated below.

```
;Timer0 ISR from good_dice.asm
;low_timer, med_timer and high_timer are three registers
;that hold elapsed time. These registers are cleared to zero
;by main program every time a key is pressed, else their value
;builds up and when it exceeds certain value, this ISR powers
;down and puts the processor to sleep
Timer_int: in save_status, SREG   ;save status in reg save_status
        cpi med_timer, $01        ;compare med_timer to 01
        brne skip_it              ;is equal then prepare to power
                                  ;down
        ldi temp, 255             ;tri-state all outputs
        out PORTB, temp           ;configure all ports as O/P
        out PORTD, temp           ;and set all values to '1'
        out DDRD, temp
```

```
             out DDRB, temp
             in temp, MCUCR         ;now set SE and SM bits in MCUCR
             ori temp, $30          ;register and make them '1' to
             out MCUCR, temp        ;select power down mode of sleep
            .sleep                  ;now sleep off
             out SREG, save_status  ;restore status
             reti                   ;return.. well it doesn't matter
  skip_it:   inc low_timer          ;if not, increment low_timer
             cp low_timer, r22      ;if it overflows to 0, then incr
             breq inc_med           ;med_timer too.
             out SREG, save_status  ;else restore status and return
             reti

  inc_med:   inc med_timer          ;incr med_timer and check if it
             cp med_timer, r22      ;overflows
             breq inc_high          ;if yes, then incr high_timer
             out SREG, save_status  ;else restore status and return
             reti

  inc_high:  inc high_timer         ;incr high_timer
             out SREG, save_status  ;restore status and return
             reti
```

10.7 Fabrication

Figure 10.8 shows the photograph of the dice circuit assembled on a general-purpose PCB. The circuit was assembled using the same general fabrication techniques presented in a previous chapter. For the AT90S1200 controller, we have used a 20-pin socket, which is a

FIGURE 10.8 Photograph of the completed dice circuit board.

FIGURE 10.9 Photograph of the solder side of the dice circuit board.

good idea for prototypes. If the processor is to be dedicated to this circuit and it is felt that the code works satisfactorily and would not need further revision, in subsequent boards the processor could be soldered directly onto the PCB to save the cost of the socket.

It is a good idea to solder the resistors, the socket, and the tiny switches right at the beginning. Later, the capacitors and the crystal are soldered and in the end, the LEDs are put in place. After all the components are soldered, proceed to wiring the components together. Figure 10.9 illustrates the solder side of the dice circuit board.

10.8   Testing

After the circuit is assembled on the general-purpose board, it is a good idea to inspect for possible shorts and open or unwanted electrical connections. A multimeter comes very handy to check for open connections and shorts in the circuit board. Check if the supply voltages are connected at the right place and the LEDs and the processor are mounted with the right polarity. Connecting the processor the other way around will lead to a lot of grief.

Once you are satisfied that all the connections are proper, insert the processor in its socket and apply power to the circuit. The applied voltage should be +5 V DC on the processor supply input pins. For this project, all the LEDs should light up immediately. Press and release the sample switch, and a random number between 1 and 6 should appear on the LEDs. Press and release this switch many times and check that the sequence in fact appears to be random. Also check that after pressing and releasing the switch, the display blanks briefly before the new number is displayed. If the circuit seems to function as described, congratulate yourself. If not, get ready for some detective work. First and foremost, check with a multimeter (in the DC voltmeter mode) that the supply voltage appears at points that it should, such as between pin 20 and ground of the processor. Next check if the reset switch is not sticking. This can be checked by monitoring the voltage on the reset pin (pin 1) of the processor. Put one of the multimeter probes at pin 1 and the other at ground (pin 10). When the RST switch is not pressed, the voltage at pin 1 should be around the supply voltage of +5 V. Now press the RST switch; the pin 1 voltage should be 0 V now. Release the RST switch and watch this voltage go up to +5 V again. If this is not happening, suspect your reset circuit composed of the resistance R4, capacitor C4, and the RST switch J1.

If the circuit seems to reset properly and is still not functioning as expected, time to look for the oscillator circuit. For the components illustrated, the circuit works at 2.000 MHz, and monitoring X2 (pin 4) should show sinusoidal oscillations at 2 MHz on an oscilloscope. If this is not happening, the culprit could be the crystal or the two capacitors C2 and C3.

Check the operation of the sample switch to see that the logic at pin PD0 changes when the switch is pressed and when released. Another possibility could be that the LEDs are arranged the other way around; check that, too, and correct that if needed. This covers the possible ways in which things could go wrong in this small project.

10.9   Usage

Well, using the dice is as simple as saying cheese. Just connect the dice to a suitable +5-V source capable of supplying a few milliamperes (25 mA or so) and you are ready to go.

To adapt the circuit to run off a battery, there are two options; either to use a +9-V box battery or use four 1.5-V cells. If you want to use a +9-V box battery you could use a 78L05 voltage regulator to get the +5-V supply voltage. The 78L05 voltage regulator comes in a small TO-92 package and can be easily put on the same PCB as the dice. However, since it requires a minimum voltage of +6.7 V to provide the +5-V output, it cannot be used if you choose to use four 1.5-V cells. Using four cells of 1.5 V each will give off +6 volts, and to get +5 V out of it you could use the LP2940 voltage regulator as described in the earlier chapter on system design. You could also choose to drop the voltage to something close to +5 V with the help of two series diodes.

Once you put the required power supply in place, just punch away at the sample switch and watch the dice roll off.

10.10   Power Consumption

As a portable utility, it will most probably be battery powered. The main concern with battery-powered devices is the power consumption, both while operating the device and when the device is not in use. The idle and power-down features of the AVR controllers come in handy in minimizing the power usage, especially when the device is not in use (Figure 10.10).

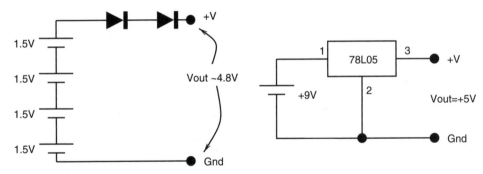

FIGURE 10.10 Possible sources of supply voltage for the dice circuit.

I measured the current consumption of the device when in use and when the device was put in powered-down sleep mode.

VCC(V)	ICC(MA)	STATE
+5	28.5	All 7 LEDs lit
+5	1 μA	Power-down sleep with all LEDs off.

However, the key to minimizing the current consumption during the power-down sleep mode is not merely activating the power-down sleep mode in software, as I found out after some efforts. The AVR ports (PORTB and PORTD in our case) if configured in high-Z floating state or as inputs, consume quite some current (about 600 μA for both the ports) even if the device is put in power-down sleep mode. To minimize this current consumption, I had to configure both the ports as outputs and I set the state of all the port bits to "1", which resulted in a 1-μA current consumption in power-down sleep mode, which is really amazing.

10.11 Adapting the Circuit to an AT90S2343

We have used a 20-pin AT90S1200 processor for this project. The project actually requires only four I/O pins—three for LED outputs and one for the sample switch input. This can be easily provided by an 8-pin AT90S2343 controller. However, the controller would need to be programmed with its internal oscillator. Any AT90S2343 could be used by programming it appropriately with the help of a parallel programmer (the serial programming methods cannot change the internal oscillator fuse bit).

The advantage of using an AT90S2343 in the internal RC oscillator enabled mode is that of reduced board space (an 8-pin DIP as opposed to a 20-pin DIP package) and component count, as it would not need the crystal and associated capacitors.

Figure 10.11 illustrates the block diagram of the dice circuit using an AT90S2343. Porting the circuit would need modifications to the software, too. The sample switch is

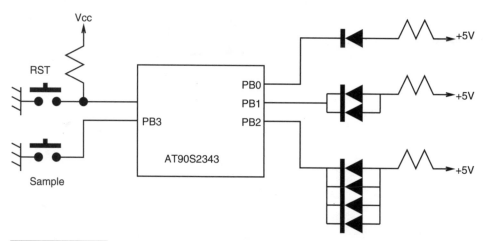

FIGURE 10.11 Block diagram for the electronic dice using an AT90S2343.

now connected to the PB3 and not PD0, as the 2343 does not have a PORTD. Secondly, the software would need to initialize the stack pointer appropriately. The AT90S1200 processor has a hardware stack that allows up to three nested subroutines. The processors in the AVR family other than the AT90S1200 use a software stack (in the internal RAM area), and so the stack pointer needs to be initialized. Once these changes are implemented, an AT90S2343 could well be used instead of the AT90S1200.

AVR PROJECT 2: A MORSE KEYER

11.1 At a Glance

Salient features of this chapter are:

1. An interesting project for amateur radio enthusiasts.
2. Uses an AT90S1200 and sundry components.
3. Discusses alternative designs.
4. Shows how to modify the circuit and code for an 8-pin AT90S2343 or ATtiny22.

11.2 Introduction

A large number of amateur radio enthusiasts all over the world employ Morse code for radio communications. Morse code is an international system of signals employed in radio telegraphy in the land-telegraph systems of all countries.

For that they must first learn Morse code. Morse code is a signaling standard using audio tones. It relies on tone duration for encoding whatever symbols you want to encode. The code specifies two types of tone durations: a short duration and a long duration. The long duration is three times the duration of the short duration tone. The duration of the tone itself is variable, and the actual value determines the speed of transmission. Symbols that are encoded comprise a sequence of short and long tones. The various tones in a symbol are spaced with periods of silence (i.e., periods of no tone). These periods are of the same duration as the short-duration tone. Symbols separated by no tone period are equal to that of a long-duration tone. These symbols are called dots and dashes. If a dot is for a time T, then the dash is for a time 3T.

A standard form of Morse code that is acceptable everywhere is called International Morse code. To learn Morse code, you need to listen to the tones for each symbol and remember them.

INTERNATIONAL MORSE CODE FOR THE ALPHABETS

A	.-	Alfa
B	-...	Bravo
C	-.-.	Charlie
D	-..	Delta
E	.	Echo
F	..-.	Foxtrot
G	--.	Golf
H	....	Hotel
I	..	India
J	.---	Juliet
K	-.-	Kilo
L	.-..	Lima
M	--	Mike
N	-.	November
O	---	Oscar
P	.--.	Papa
Q	--.-	Quebec
R	.-.	Romeo
S	...	Sierra
T	-	Tango
U	..-	Uniform
V	...-	Victor
W	.--	Whiskey
X	-..-	X-ray
Y	-.--	Yankee
Z	--..	Zulu

INTERNATIONAL MORSE CODES FOR NUMBERS AND PUNCTUATION

1	.----	Full-stop (period)	.-.-.-
2	..---	Comma	--..--
3	...--	Colon	---...
4	....-	Question mark (query)	..--..
5	.....	Apostrophe	.----.
6	-....	Hyphen	-....-
7	--...	Fraction bar	-..-.
8	---..	Brackets (parentheses)	-.--.-
9	----.	Quotation marks	.-..-.
0	-----		

One of the common ways to learn Morse code is to buy recorded audio cassettes which play sounds of all the symbols. You hear this again and again till you memorize all the symbols. Then you want to be able to generate Morse code on your own before you start using it for radio transmission.

In this chapter we will build a simple yet powerful Morse code generator. A very simple Morse code generator could be built using a simple buzzer in series with a Morse key. However, for radio transmission, that is not suitable. Also, these days it is more common to use a paddle rather than a Morse Key. A paddle is a mechanical device that has two sets of contacts. In one position, it generates dashes and in the other position, it generates dots. Our Morse code keyer is designed to work with a paddle type of Morse key.

11.3   Design Specification

For this circuit, we needed a simple circuit that could be connected to the popular More Paddle if available or even to ordinary push-button type switches for generating the Morse dots and dashes. It was also desired, as always, to keep the current consumption as small as possible to facilitate extended battery operation. It was also felt necessary to be able to control the Morse code generation speed, and above all, to keep the circuit small so as to be portable.

11.4   Design Description

To meet the above design objectives, we came up with a simple AVR AT90S1200-based Morse keyer circuit. Figure 11.1 illustrates the block diagram of the Morse Keyer. It is a very simple circuit and needs only a few switches, an LED, a buzzer, and few other sundry

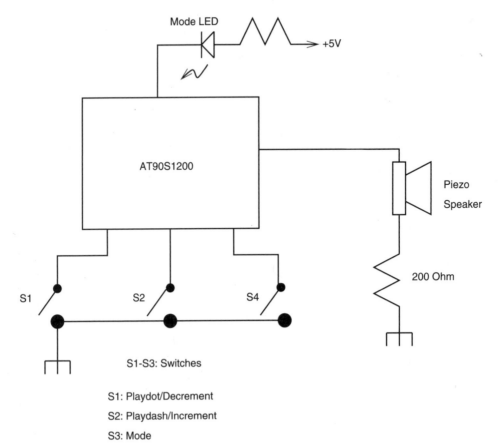

FIGURE 11.1 **Block diagram of the Morse Keyer.**

components to build, besides the AVR processor. I decided to use the AT90S1200 proces-sor, the simplest processor from the AVR family, as this project did not need any more fea-tures than provided by the AT90S1200.

Figure 11.2 illustrates the circuit diagram of the Morse Keyer. A small piezo buzzer was used to provide the audio tone. The was operated at a clock frequency of 4 MHz using an external crystal. To reduce the component count further, one could use the internal RC oscillator of the 1200 processor. To use an ordinary 1200 processor in the internal RC oscillator mode, you need to enable the on-chip oscillator by clearing the RCEN control bit to "0." This chip is shipped with this bit set to "1" and can be cleared to "0" using the parallel programming mode only. However, if you have the AT90S1200A part number, it is shipped with the RCEN bit cleared to "0" and ready to use the internal RC oscillator as the processor clock source.

The design has three push-button switches labeled mode, playdot/decrement and play-dash/increment. Two of these switches are dual purpose. They are used to decrement and increment Morse code generation speed in one mode and as dot and dash tone generator in the other mode. When the Mode switch is pressed, the other two switches are used to adjust the speed. When the Mode switch is released, these switches function in their other

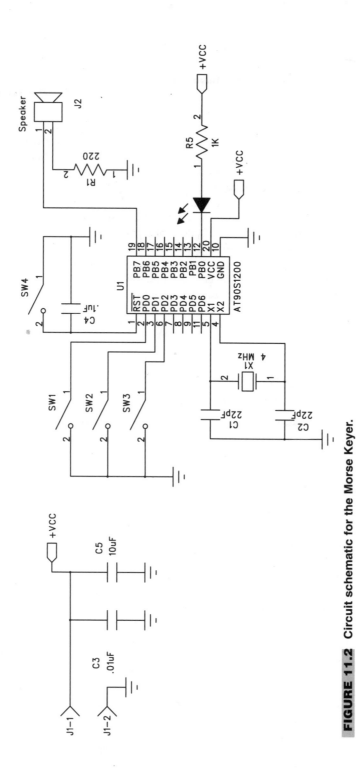

FIGURE 11.2 Circuit schematic for the Morse Keyer.

mode for generating the dots and dashes. The LED indicates the mode of operation of the circuit. When the Mode switch is pressed, the LED is ON, indicating that the other two switches are in the speed increment/decrement mode.

To generate dots, the playdot/decrement switch is pressed (i.e., the logic at this switch is "0"). This generates an audio tone of a duration of a dot. If this switch is kept pressed, the dots repeat with an intervening period of no tone equal to the dot period. Similarly, if the playdash/increment switch is pressed, an audio tone for the period of a dash is generated, and when keeping this switch pressed, continuous dashes are generated, with intervening periods of no tone equal to a dot period.

By manipulating the two switches, tone sequence for any switch can be produced. The main heart of the software consists of Timer0 interrupt, which is used to generate the audio tone by toggling the PB7 pin. A piezo buzzer is connected to this pin.

The Timer0 is programmed to generate an interrupt every 1.12 ms (i.e., an interrupt frequency of 892 Hz). Since the Timer0 ISR toggles the PB7 bit, the resulting audio tone has a frequency of 446 Hz. The speed of generation of the Morse code is implemented by counting the number of audio tone pulses generated. A software counter is used to monitor the number of pulses generated. If the number of pulses generated for a dot (or a dash) equals this count, then the generation is disabled for a period equal to a Dot time. The speed can be adjusted between 5 words per minute to more than 40 words per minute.

11.5 Possible Alternatives

Since the total I/O requirement as illustrated in Figure 11.2 is just five I/O lines, it is possible to use the AT90S2343 or the Tiny22 processor. Either of these processors could be used for this project by operating the processor in the internal RC oscillator clock mode.

11.6 Fabrication

This circuit was tested on the Atmel AVR evaluation kit and no separate PCB was used.

11.7 Design Code

The design for the project is available in the code directory in the file morse1.asm. The design code is split up in small subroutines as follows:

1. Initialization: This section of the program is executed when the power is first applied to the processor or the reset switch is pressed. In this section, the code initializes the various registers, the timer, port pins, etc.
2. Loop Here: This is the main part of the code. In this part, the processor checks if a key has been pressed and whether the Mode key is also pressed. Depending upon which key combination has been pressed, appropriate action is taken. The possible combinations are to execute a subroutine to increase the Morse code generation speed, decrease

Morse code generation speed, to generate a Dash, or to generate a Dot.

3. Get Keycode: This routine is called to check which key has been pressed. If no key is pressed, this routine waits for a key or a combination of keys to be pressed. It then returns a key code for the main program to interpret and take appropriate action.

4. Play Dash: This routine generates an audio tone for a period equal to three times a dot period. After the tone, it produces a period of no tone equal to a dot-period length.

5. Play Dot: This routine sets up the Timer0 interrupt routine to generate an audio tone for T time units. This is the basic time period of the Morse code generation speed. After the tone, it produces a no-tone period of T time units.

6. Increment Speed: This routine increments the speed counter in increments of 10.

7. Decrement Speed: This routine decrements the speed counter in steps of 10.

8. Timer0 ISR: This is the heart of the program. This routine generates the audio tone at a frequency of 446 Hz. It also increments a temporary counter to indicate how many pulses of the audio tone have been generated. This counter is used by the play dash and the play dot routines to determine if the dash (or dot) period is over or not.

It should be noted that the AT90S1200 does not have an SRAM-based stack, but a hardware stack that has a maximum depth of three. Therefore the code is written in such a way that at any given time there are no more than two nested subroutines. This is to accommodate the possibility of the Timer0 interrupt occurring when up to two nested subroutines are called from the main program.

11.8   Testing the System

After the code was written on the Atmel AVR evaluation board, pressing the keys gave a response as expected. The speed could be varied by pressing the Mode switch in combination with the playdot/decrease speed or the playdash/increase speed keys. I then observed the Audio output on the PB7 pin of the processor on a digital oscilloscope. The traces captured on the scope are illustrated in Figures 11.3, 11.4, and 11.5, and they confirm the correct operation of the system.

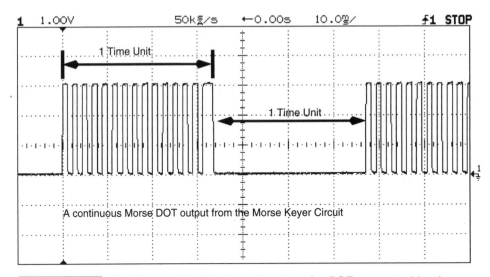

FIGURE 11.3   Oscillogram for Morse code output for DOT, generated by the Keyer circuit.

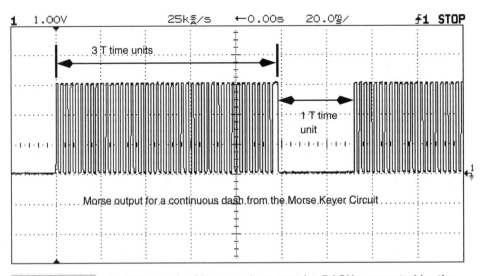

FIGURE 11.4   Oscillogram for Morse code output for DASH, generated by the Keyer circuit.

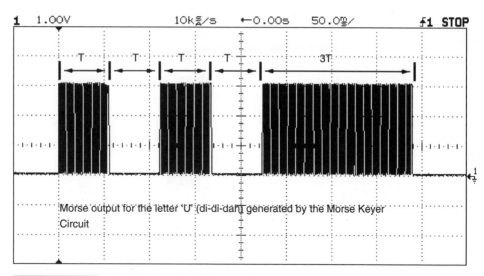

FIGURE 11.5 Oscillogram for Morse code output for the character U, generated by the Keyer circuit.

AVR PROJECT 3: A SIMPLE
DUAL-CHANNEL VOLTMETER

12.1   At a Glance

In this chapter we look at an AT90S2313-based dual-channel voltmeter using the MAX111 ADC. The features of the system are:

1. Dual-channel voltmeter with LCD display.
2. Driver software in C.
3. User interface switches.
4. Can be modified to download readings to a PC.

12.2   Introduction

In this chapter we have a project that connects the AT90S2313 controller to the MAX111 ADC and a 2 × 16 character LCD display. The controller converts voltage on both channels of the ADC and displays the result on the LCD. The voltage corresponding to channel 1 is displayed on one of the lines, and the voltage corresponding to the other channel is displayed on the other line. The voltmeter indicates the sign of the applied voltage as well as provides an overvoltage indication. Figure 12.1 illustrates the voltmeter block diagram.

12.3 Design Description

Figure 12.2 illustrates the circuit schematic for the dual-channel voltmeter. The LCD is connected to PORTB pins with a 4-bit interface in write-only mode. The ADC is connected to the PORTD pins. Switches SW2 and SW3 have been provided for user interface. Currently, they are not being used but can be utilized for any specific need, or they could be replaced and the PD0, PD1 pins could be used to communicate with a serial port for data-logging purposes.

The reference voltage to the ADC is through a 2.5-V zener. An LM336-2.5 V could be used. The range of the voltmeter is ±2.5 V maximum. By changing the reference voltage, the input range of the voltmeter can be changed. The circuit needs +5 V for operation. The controller operates at 3.58-MHz clock.

12.4 Usage

This dual-channel voltmeter can be used when there is a need to monitor two voltages simultaneously.

12.5 Fabrication

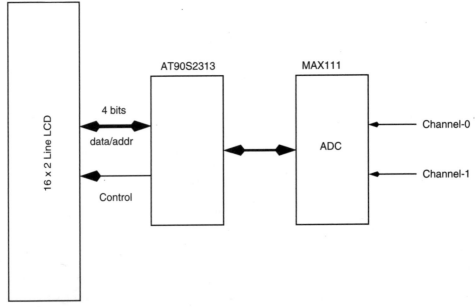

FIGURE 12.1 Block diagram for the dual-channel voltmeter with LCD.

The voltmeter was fabricated on a general-purpose PCB about 2 inches a side, and the completed circuit board is illustrated in Figure 12.3.

12.6   Design Code

The code for this project was written in C using the SPJ systems' C-AVR compiler and is available in the code directory in the file adc.c. Figure 12.4 illustrates the MAX111 ADC signals as captured on a logic analyzer.

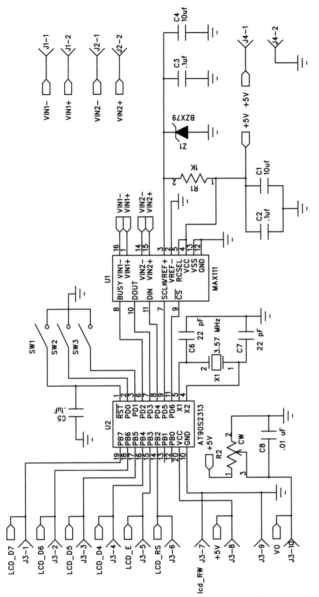

FIGURE 12.2 Circuit schematic for a the dual-channel voltmeter with an LCD display.

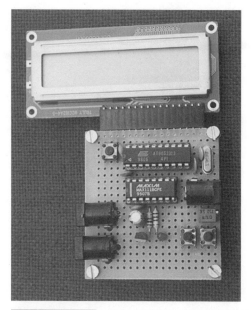

FIGURE 12.3 Photograph of the dual-channel voltmeter.

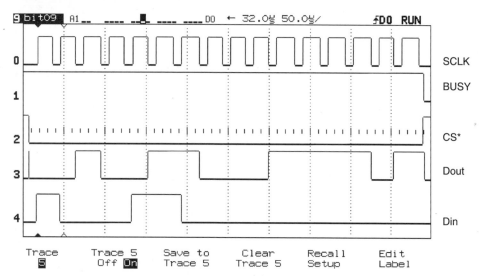

FIGURE 12.4 Logic analyzer screen capture of the MAX111 ADC readout by the AT90S2313 controller.

13

AVR PROJECT 4:
THE UBIQUITOUS KITCHEN TIMER

13.1   At a Glance

In this chapter we look at another simple design that uses an AT90S1200. The features of the timer are:

1. A no-frills multipurpose timer
2. Uses two 1.5-V AAA cells
3. Simple user interface
4. Input time settings with thumb-wheel switches
5. Timer is armed with a single switch
6. Shows how to use an interrupt
7. Driver code in Assembler

13.2 Introduction

Time is money. In the kitchen, knowing time could mean you have an unburnt, properly cooked dish (whether it tastes good is another matter).

Toward that end, we have a small and useful project that will allow you to keep good time. Besides in the kitchen, you could use it elsewhere, as a siesta alarm for example.

This design of an interval timer is kept very simple with no frills. Perhaps it doesn't get simpler than this design. Figure 13.1 illustrates the block diagram of the design.

Thumb-wheel switches are used to enter the time in minutes (nothing cooks in seconds anyway!). Once you set the required time, just reset the timer and it starts counting. Once the set time has elapsed, the buzzer goes on.

13.3 Design Description

The design of the timer is very simple. Figure 13.2 illustrates the circuit schematic. Only a handful of components are required to build the Timer. To keep the system simple, we have used thumb-wheel switches to input the time. More fancy solutions involving a keypad and an LCD display are also possible, but that leads to a bigger circuit and one which consumes more current.

The thumb-wheel switches are connected to the PORTB pins. Each thumb-wheel switch requires four pins to read. So the entire PORTB is used up. The PORTB is programmed as an input port with the pull-up resistors on. This means that external resistors are not required.

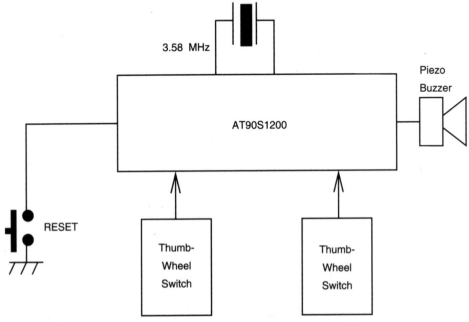

FIGURE 13.1 Block diagram of the simple kitchen timer.

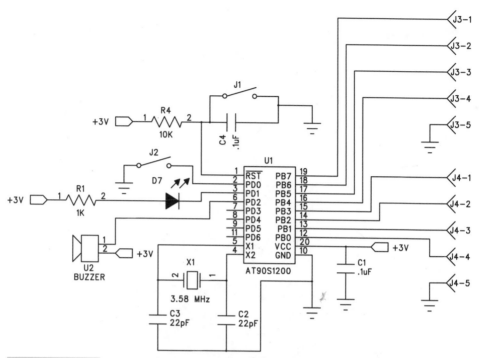

FIGURE 13.2 Circuit schematic for the kitchen timer.

The LED and the buzzer are put on the PORTD pins. The LED is connected to sink the current into the PD1 pin. The buzzer is connected to the PD2 pin. The buzzer is also set up to sink current into the pin.

The circuit operates at 3.58 MHz. The internal RC oscillator cannot be used instead of the external crystal, as the RC oscillator frequency is dependent on the supply voltage and would render the timer quite useless. For other applications where timing variation is not crucial, it is an attractive proposition.

13.4   Possible Alternatives

The design cannot fit on any smaller AVR processors, and amongst the 20-pin AVR processors, the AT90S1200 is the one with the least amount of on-chip resources. The AT90S1200 processor also consumes the least amount of current. For this application, one could reduce the crystal frequency further to reduce the current consumption even more.

13.5   Fabrication

The initial idea was tested on the Atmel's AVR evaluation board, and later the circuit was built on a general-purpose PCB. The circuit has been used in my kitchen for some time now. Figure 13.3 is the photograph of a completed circuit board. The timer board is connected to a pair of thumb-wheel switches illustrated in Figure 13.4.

13.6   Design Code

The code for this project is available in the code directory in the file kitchen.asm.

13.7   Testing

Keep an eye on all connections. In my prototype, I forgot to ground the reset capacitor and found that the circuit would reset on its own quite frequently. After much contemplation, the trouble was traced to the open reset capacitor problem.

FIGURE 13.3 Photograph of the kitchen timer.

FIGURE 13.4 Photograph of a pair of thumb-wheel switches

AVR PROJECT 5:
RADIO BEACON CONTROLLER

14.1   At a Glance

In this chapter we consider the following issues:

1. What is a radio beacon?
2. What is a radio beacon used for?
3. How to build a radio beacon controller.
4. Details of an efficient and compact single-chip radio beacon controller.
5. How to adapt the controller for your application.

14.2   Introduction

Merriam Webster's Collegiate Dictionary defines a beacon as a signal fire or other signal commonly on a hill, tower, or pole for guidance. A radio beacon is a radio transmitter emitting signals for guidance of an aircraft. However, a radio beacon is used for more than aircraft guidance.

 A radio beacon is usually put on scientific balloons, rockets, etc., for identification as well as help in locating the object (i.e., the balloon or the rocket). Usually the beacon outputs a short message over and over again, which drives an appropriate radio transmitter. In amateur radio, radio beacons have been used for quite some time. These beacons usually transmit Morse code at 22 words per minute, transmitting their call sign in 10-dB power, which is very useful for the purpose of "S" meter calibration and for judging the band conditions. These beacons usually transmit every three minutes.

A radio beacon consists of a radio transmitter capable of transmitting at the required radio frequency and required output power. The transmitter is driven by a suitable controller that stores the guidance message. Since the radio beacon is located on unattended sites, an important consideration is to optimize available power in transmitting the guidance signal. A low-power controller that consumes as little power as possible is desirable.

Figure 14.1 illustrates the block diagram of a radio beacon. The controller for a beacon can be built using a microcontroller for programmability and low-power operation.

14.3   Design Specifications

We are looking for the design of a low-power programmable controller for a radio beacon. The controller will be able to output a stored message in the form of Morse code, which will be repeated every three minutes. The repetition rate should be programmable, and it should be able to easily change the stored message. The controller should be as small as possible and should implement power-down modes of operation to minimize power consumption. The controller should work off a wide range of supply voltage. Let us see how our AVR-processor-based design meets these specifications.

14.4   Design Description

The primary requirement for a radio beacon controller is to be able to operate on as little power as possible. The controller is active for a very short time in a whole period. If we decide to operate the beacon every three minutes, then the beacon will transmit the signal for a few seconds, and after that it will be passive. Therefore, to have a controller that is

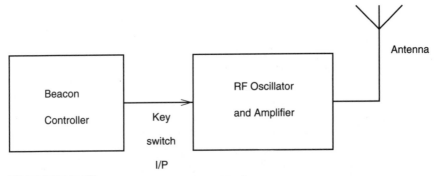

FIGURE 14.1 Block diagram of a radio beacon.

continually executing a program even when the beacon is not transmitting any signal is wasteful.

A microcontroller (with power-down features) as the controller can be operated with many of the available power-saving features. AVR processors offer two such power-saving states that reduce power consumption compared to the active state. However, the processor is in some state of inactivity during these power-saving modes and is not executing any program in either of these states.

The power-saving modes of operation in the AVR family of processors are called the idle mode and the power-down mode. In the idle mode, the CPU stops executing programs, but the timers, interrupts, and the oscillator keep working. In this mode, an external or an interrupt source of interrupt will wake up the CPU and normal program execution can resume. In the power-down mode of operation, the clock oscillator is also stopped, besides the internal timers and analog comparator. The user can select whether to keep the watchdog timer working. If the watchdog timer is enabled, then the watchdog timer can reset the processor after the watchdog timer expires. Otherwise, only an external reset or an external-level-triggered interrupt can wake up the CPU. The power-down mode of operation reduces the power consumption the most as the internal clock oscillator as well as the timers stop working. In both of these modes of operation, the internal SRAM contents are retained.

However, if one chooses the power-down mode of operation and the watchdog timer for wakeup, there has to be some means of keeping a count of the elapsed time. If your particular application needs to do something at a rate that can be met by the watchdog timer, then it poses no problem at all. The watchdog timer has a wakeup range between 15 ms and 2 s. So if your application needs to do something every 2 s approximately, then the watchdog timer could be armed to reset the processor 2 s after it has been enabled and armed. However, what if the application needs a cycle time of 20 s? In this case a count of the number of watchdog resets could be maintained in a software counter (in the SRAM), and when this count becomes 10, then the required activity could be performed. However, there has to be some means to distinguish between a power-on reset (or the external reset) and a watchdog reset.

There is a complex way to handle this situation. It works on the assumption (and high probability) that the internal SRAM locations would not have a particular sequence of numbers at power on; e.g., the chance of three consecutive SRAM locations at address $00, $01, and $02 to be initialized to say $55, $AA, and $55 respectively is very, very small. In fact, the probability that after power on, these locations will have the exact sequence of numbers is 1/16777216 or about 1 in 100 million. Thus the software could initialize these SRAM locations to the required sequence if after reset these locations are found to contain some other number, thus indicating that the reset was caused by an external reset of a power-on reset and at the same time a software counter in SRAM could be initialized to $00. On the other hand, if the reset has been due to the watchdog timer (and the SRAM locations have been initialized to the required sequence after a power-on reset or an external reset), the SRAM sequence would match and then the software would just examine the software counter, and if the required count has been accumulated in the counter, it would perform the task, reset the counter to $00, enable the watchdog timer once again, and go in to power-down mode. On the other hand, if the software counter has not reached the required count, it will just increment the counter, enable the watchdog timer, and go into power-down mode.

However, some AVR processors are equipped with some features to distinguish between a power-on reset (or an external reset) and a watchdog reset. Tiny22 is one such processor that allows a program to determine the source of reset in a simple manner.

The MCU status register (MCUSR) provides information about the source of reset. The MCUSR contains 2 bits which indicate the source for the reset as per Table 3.4 (in Chapter 3). The MCUSR register contains the EXTRF and the PORF flags. When both of these flags are "0," the reset source is the watchdog reset, and when either PORF is "1" or the EXTRF is "1," the source of reset is power-on reset of the external reset (i.e., the reset pin is grounded momentarily).

By using this elegant feature of the processor, one could implement a scheme to provide a large (and programmable) time period between generating the beacon signal repeatedly.

The other design goal is to be able to store any Morse code sequence in the processor and be able to easily change that when required. We could either program the message as a string in the flash program memory or in the EEPROM, since the Tiny22 has 128 bytes of EEPROM. One advantage of storing the Morse code in the EEPROM is that the message could be changed on the fly without having to program the flash program memory. Therefore we decided to store the Morse code in the EEPROM of the controller rather than in the program memory (Figure 14.2).

This beacon controller design is a truly one-chip solution with just a power-on-reset circuit. No crystal is required, as we will be using a Tiny22 part which has an internal RC oscillator for the clock. The output of the controller controls the transmission of a suitable RF oscillator/amplifier.

Figure 14.3 illustrates the block diagram of our Tiny22-based beacon controller. One of the outputs is used to generate the audio sidetone of the Morse code signal, and another output is used to switch the RF transmitter on and off (Figure 14.4).

Figure 14.5 illustrates the circuit schematic for the beacon controller. A reset switch is provided to reset the system manually, and an LED is a visual indicator when the system starts transmitting the Morse signal.

The software could be programmed to transmit every few minutes. We could choose to transmit a five-WPM (words per minute) message from the controller, and this could be repeated every minute or less (say 30 seconds). It is common to transmit carrier for a few seconds followed by the message. Let us choose two seconds of CW, two seconds of

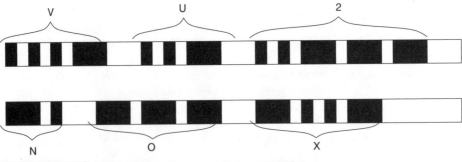

FIGURE 14.2 Morse output for my callsign VU2NOX.

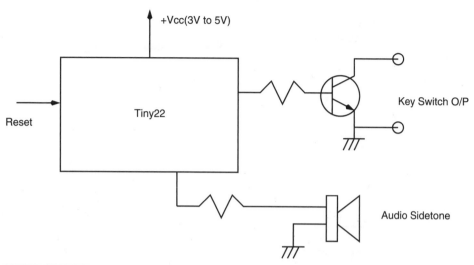

FIGURE 14.3 Block diagram of a radio beacon controller using the Tiny22 processor.

silence followed by the message, and this sequence is repeated. The software could be changed to choose any Morse transmission speed between five WPM and thirty WPM.

This simply means that one of the I/O pins is set to 1 for transmission of the carrier (by enabling the RF transmitter) and set to 0 for disabling the carrier. The Morse code characters, a dot and a dash, have 1:3 time relation, i.e., a dot is a sound for 1 time unit and a dash is a sound for 3 time units. The interval between a dot and a dash of a single character is 1 time unit, e.g., code for V = dit dit dit dash, which is 1 0 1 0 1 0 1 1 1. Here a 1 means that the audio tone is on and 0 means that the audio tone is off. Three consecutive 1's mean that the sound is on for 3 time units. Space between any two characters of a word is 3 time units of silence and between two words is 5 time units. So a complete transmission of my call sign, VU2NOX will be as illustrated in Figure 14.2.

The Tiny22 has a watchdog timer that, if enabled, resets the processor when it expires. The timer has its own oscillator with a nominal frequency of 1 MHz. The watchdog timer can be set so as to expire in 16-K clock cycles to 2-M clock cycles. This gives a watchdog time span ranging from 16-ms minimum to 2-s maximum.

I have chosen to use the watchdog timer to implement the wake up from power-down sleep state of the processor. The watchdog timer is programmed to expire (and reset the processor) after 2 s. For this project, the goal is to transmit the message every 30 seconds. To do that, I use an internal RAM location (the internal RAM is not initialized at RESET). If the contents of this location exceed 15, I execute the routine to transmit the message. Else, the RAM contents are incremented by one and the watchdog timer is enabled to reset the system after 2 s. After the watchdog timer is armed and the processor executes the sleep instruction, the processor enters the power-down mode. It wakes up again when the watchdog timer expires and resets the system. This scheme allows system operation at minimum power consumption. Figure 14.4 illustrates the flowchart for the beacon controller system code.

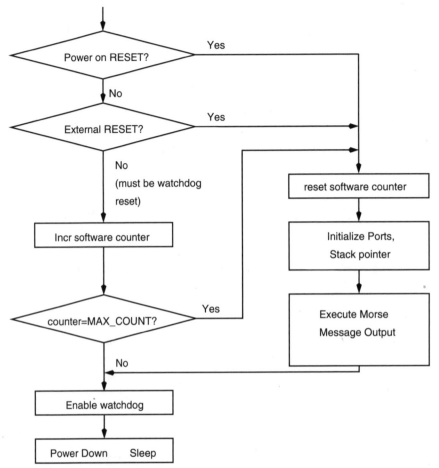

FIGURE 14.4 Flowchart for the beacon controller program.

14.5 Fabrication

The system was fabricated on a general-purpose PCB as illustrated in Figure 14.6. Since the whole system is very small, fabricating the circuit was quick and easy.

14.6 Design Code

The design code for the beacon controller was developed using the flowchart illustrated in Figure 14.4. The code was split in small subroutines. Initially, the system code to check the watchdog reset and to distinguish the watchdog reset from power-on reset was written and tested. The Morse code generation subroutine was tested separately and then integrated into the main program. A table that encodes the Morse code was created and stored as program memory data and stored in the flash program memory. The actual message was

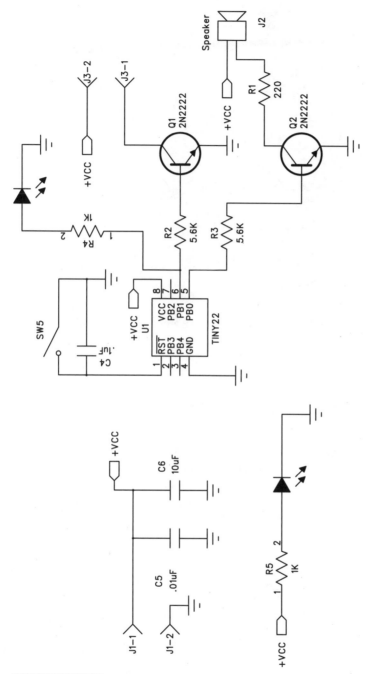

FIGURE 14.5 Circuit schematic for the radio beacon controller.

FIGURE 14.6 Photograph of the beacon circuit board.

stored in the EEPROM as index into the Morse table. Since the EEPROM is 128 bytes, a message of up to 128 characters can be stored and generated by this system. The code for this project is available in the code directory in the file mtutor1.asm.

14.7 Testing

The system was tested using standard test equipment. One easy test was the fact that the system could generate correct audio tone for the stored message. Figure 14.7 shows the Morse audio side tone signal and RF oscillator key switch output generated by the controller.

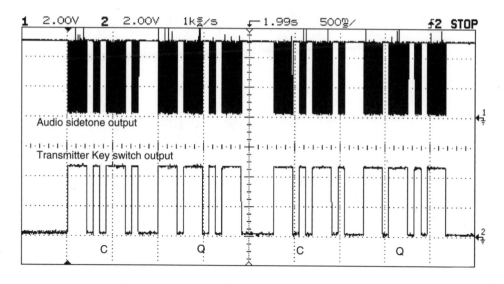

Audio sidetone output

Transmitter Key switch output

FIGURE 14.7 Scope trace for the audio sidetone as well as the transmitter key switch output generated by the beacon controller. The trace shows four Morse codes for the characters C Q C Q.

AVR PROJECT 6: ASTRODAT: A STAND-ALONE DATA ACQUISITION SYSTEM

15.1 At a Glance

All about data acquisition systems

1. Describe a matchbox-size low-power DAS using only 3 ICs
2. A complete DAS for astronomical applications
3. An OS-independent readout using an RS-232 port

15.2 Introduction

There are occasions when it is necessary to record data in an unattended manner over extended periods of time.[1] Such requirements can often be met with a suitable data acquisition system connected to a PC. Often enough, there are occasions to log such data in remote wilderness with no access to suitable power. Such requirement can be met with an autonomous data acquisition system that runs off battery power.

---

[1]As Ambrose Bierce might have said, "The code presented in this chapter was developed by Saurabh Jain and Smita Mohan and to whom is rightly due the credit for the merit that it may have."

Figure 15.1 illustrates the block diagram of a PC-hosted data acquisition system. The data acquisition hardware contains suitable electronics front-end circuitry to digitize the input analog data. The converted digital data is uplinked to the PC through the connecting link between the PC and the data acquisition hardware. The link itself could be serial (RS-232, RS-485, USB, IrDa) or parallel (parallel port, ISA expansion card). The PC software would acquire, store, and eventually process the acquired data.

As mentioned above, the problem with this setup lies in meeting the power requirements for running the PC and the data acquisition hardware in remote locations—not to mention the security needs of such a setup. To some extent the power requirements could perhaps be solved by using a Notebook PC, but not for extended periods of time.

When it comes to low power, extended-period-acquisition applications, nothing beats the setup illustrated in Figure 15.2. The controller is armed to acquire data in the required format (which includes such information as the sampling rate, etc.) and then taken to the site where the acquisition takes place. Upon completion of the acquisition activity, it is brought back to civilization and the stored data is read out to a PC for analysis.

In this chapter we look at a couple of such data acquisition system designs. The next section describes a simple paper design using an 1-channel 12-bit ADC and serial EEPROM for data storage. EEPROMs are available in 64-Kbyte capacity in 8-pin DIP packages, and up to 4 of these can be cascaded to give 256-Kbyte storage. The data is stored in the EEPROM and can be read out through the PC parallel port in a novel way.

Using EEPROM has an operational advantage: It can retain data even in the absence of power. However, there is a caveat: EEPROMs cannot be written as fast as conventional SRAM, and this is a disadvantage that one has to live with. It is, however, possible to alleviate this problem to some extent by employing buffer memory, but again at a cost of increasing system complexity or increasing the number of EEPROMS and striping the data storage across the EEPROMs.

Later, I describe a complete and tested DAS that is specifically designed for use in astronomical applications. It can also be used elsewhere without any change.

However, the keyword in the design of both the systems is simplicity. Both of the designs have a single and critical design objective:

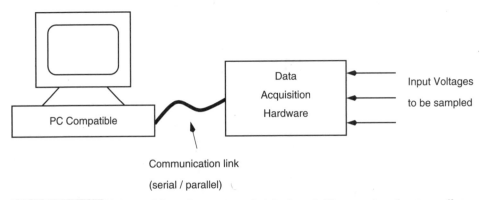

FIGURE 15.1 Using a PC and an external data acquisition system for recording data.

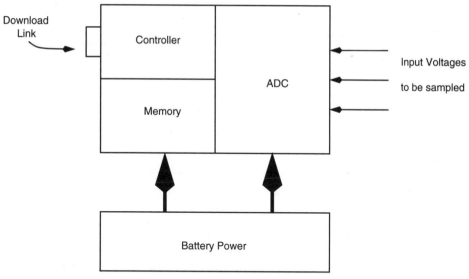

FIGURE 15.2 An autonomous data acquisition system.

1. Small size with minimum power consumption.

Keeping in mind the limited I/O offered by single-chip microcontrollers, we have gone in for components that have simple serial interfaces. Usually, memory devices such as EEPROMs are available with the two-wire I2C interface, while ADCs are available with the SPI/Microwire (which requires three wires) interface. It would be nice to have ADCs with IIC interfaces as well, but unfortunately, there aren't many ADCs with IIC interfaces.

15.3   Design Description for the SniffStick

Figure 15.3 illustrates the block diagram of SniffStick, our low-power data acquisition system. It uses an AT90S2343 (or Tiny22) processor in an internal RC oscillator mode. It takes all five I/O port lines from the controller to connect all the peripherals. A docking port with connection to the controller reset is used to configure the system prior to data acquisition, and after the data is acquired, the docking port is used to read out the data. The docking port is connected to the PC parallel port. The PC parallel port holds the controller under reset condition and takes control of the SDA and SCL lines of the AT24C512 serial EEPROM and reads out the entire memory (Figure 15.4).

The operation of the SniffStick is best understood by breaking up the design in three parts:

1. Interface to the ADC
2. Interface to the EEPROM
3. Interface to the docking port

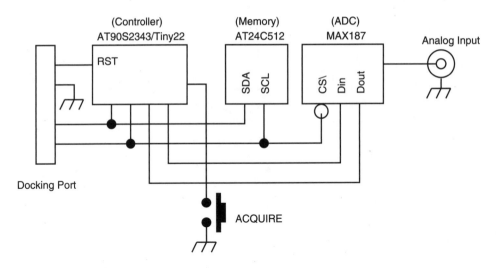

Docking Port is connected to the Host PC for configuring the SniffStick
as well as for data readout

FIGURE 15.3 Block diagram of SniffStick.

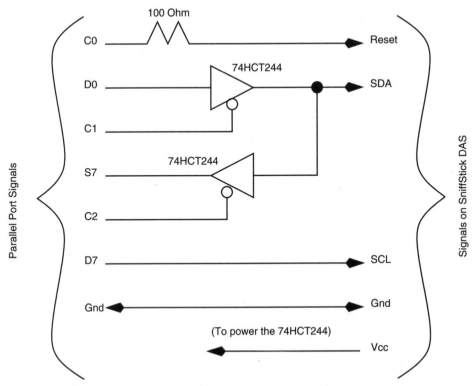

FIGURE 15.4 Block diagram of the PC parallel-port-based docking port for the
SniffStick DAS.

Let us see how this design works. The ADC I have chosen for this system is a tiny 8-pin, 12-bit ADC MAX187 from Maxim. The ADC performs a 12-bit conversion in about 10 μs. The ADC offers a serial three-wire SPI/MICROWIRE interface. We have connected three I/O lines from our processor to the ADC. One of the signal lines to the ADC, the CS* line, has been shared with the SCL signal of the EEPROM.

The serial EEPROM has an IIC interface and thus requires two I/O lines from the processor, one of the signals being shared with the CS* signal of the ADC. This arrangement is possible as the shared signal between the ADC and the EEPROM requires complementary logic signals for each device, i.e., when this signal line is "1," the EEPROM is active and the ADC is inactive, and when this signal is "0," the EEP-ROM is inactive (provided that the SDA signal is held to "1" by the processor) but the ADC gets selected.

A switch is connected to the available I/O line for triggering an acquisition cycle. Figure 15.5 illustrates the circuit schematic for the SniffStick, and Figure 15.6 is a photograph of the SniffStick circuit board.

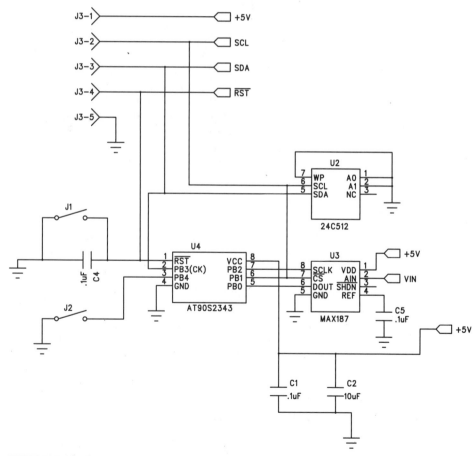

FIGURE 15.5 Circuit schematic for the SniffStick.

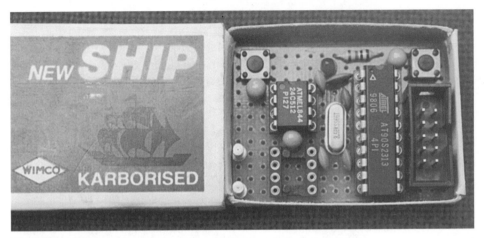

FIGURE 15.6 Photograph of the SniffStick under fabrication.

15.4   Using the SniffStick

How is the SniffStick intended to be used? All data acquisitions need to acquire a data at a certain rate. The processor for SniffStick is clocked at the internal RC oscillator clock, which has a nominal value of 1 MHz at 5 V. The EEPROM can only be written a byte every 10 ms unless one writes the EEPROM in page mode in which a whole page of memory can be written in 10 ms, which could increase the average byte write speed.

The SniffStick is configured in such a way that the EEPROM is not only used to store the acquisition data but also the information about the data acquisition rate. I choose to reserve the first two locations in the EEPROM (at address $0000 hex and $0001 hex) for an acquisition time multiplier. This 16-bit number is used to multiply the basic rate of 10 ms to achieve any desired rate. Let us see how this works. Suppose you want to acquire data at the rate of 200 ms, then

$$(TimeMultiplier) * 10ms = 200ms$$

therefore,

$$TimeMultiplier = 20 = 14 \ (hex) :$$

Thus the first two bytes in the EEPROM are stored with the following numbers:

Address $00: Data $00

Address $01: Data $14

These two numbers are programmed into the EEPROM through the docking port at the time of system configuration at the base station. After that, the SniffStick can be ported to the field and used rightaway. To start acquisition, the Acquire key is pressed. The program will then read the first two locations of the EEPROM to determine the acquisition rate and program the internal timer appropriately. After that the data from the ADC is acquired and stored in the EEPROM at locations starting at address $0002. Each sample is two bytes wide (to accommodate the 12-bit ADC result). Between acquisition cycles, the processor goes in sleep mode to conserve power. After the entire memory has been filled up, the processor goes into sleep mode and can be now brought back to the base station and the stored data read out through the Docking port. The next section describes a complete DAS on similar principles, but with some modifications to accommodate more features.

15.5   AstroDat: A Complete DAS for Astronomical Application

The SniffStick DAS is an interesting and useful piece of hardware. However, for certain astronomical data acquisition applications, we needed something more than SniffStick could offer, namely a time-stamping facility as well as the ability to store multiple observation data in a single memory device.

The particular astronomical observation that I am referring to is photometric data logging. A photometer is an instrument that converts light to voltage. Astronomical photometers are low-noise devices that are capable of detecting faint light intensities emanating from stars and other objects in the sky.

Typically, for an astronomical data acquisition run, you would like to record data related to many different stars and objects in a single night of observation. Besides, you would also like to record the ambient night sky brightness as a reference.

This means that the date-recording system should have the ability to start and stop a particular observation and maintain these different sets, which could be separated out later.

To accommodate this feature, I modified the basic SniffStick design to incorporate an RTC and an extra switch. To do this I had to migrate to a bigger processor with more I/O lines. Figure 15.7 illustrates the new design.

The data storage was implemented with a new structure, which is as follows:

1. The unit of data storage in the EEPROM was decided to be 16-bit word.
2. Data stored in the EEPROM has tags to help interpret the data. Each set of observations is identified by a unique "set number" tag. Each set of observations will have a start tag (containing the set number) at the beginning of the set and a stop tag indicating the end of the set.
3. Each set of data has a start tag followed by a time tag, which indicates the time at the start of the set, and this is followed by the actual data. There can be as many data points as possible, limited by the EEPROM storage capacity.

15.6   AstroDat User Interface

The user interacts with the AstroDat system through the three switches and an activity LED on the PCB. The switches, as illustrated in Figure 15.7, are: Reset switch, which can

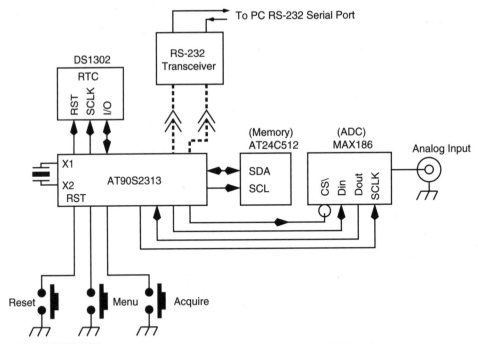

FIGURE 15.7 Block diagram of the AstroDat Data Acquisition System.

be pressed at any time to restart the system; the "Menu" key, which can be pressed after the system reset to interact with the system for configuration and data readout through a PC serial port; and the "Acquire" key, which can be pressed after the system reset to begin acquisition. The activity LED on the board indicates when the system is busy acquiring data. When the LED is ON, the system is acquiring data as per the sampling rate and when the LED is OFF, the system is waiting for either the "Acquire" key or the "Menu" key. Pressing the "Acquire" key starts a new set of data to be acquired. Pressing the "Acquire" key when the LED in ON terminates the ongoing data acquisition process and completes the current set.

The AstroDat DAS offers an interactive user interface on a PC terminal emulator program through its serial port. The DAS can be programmed for various functions through this user interface. The AstroDat hardware communicates on a serial port set at 9600 bps, 8 data bits, 1 stop bit, and no parity bit. The functions are:

1. Initialize RTC. Allows the user to set the RTC time.
2. Read RTC. Allows the user to verify the time.
3. Read EEPROM. Allows the user to read the EEPROM contents online in small chunks.
4. Enter delay. This is a 2-byte value that sets the interval between two data points.
5. Read delay. Allows the user to read the delay value.
6. Write to file. This allows the user to download the entire EEPROM contents that the user can pipe into a file for further analysis.
7. Exit Menu. Required to close the session.

15.7   Design Description

Figure 15.7 illustrates the block diagram of the AstroDat Hardware. The hardware is fairly similar to the SniffStick, except for additional components such as the RTC and an extra switch. Besides, the AstroDat communicates to a host on a serial port rather than the Docking port as in SniffStick. This is because the AstroDat Uses a bigger 20-pin AT90S2313 processor with built-in serial port.

Figure 15.8 is the circuit schematic for AstroDat. The system uses AT90S2313 at 3.58 MHz. Any other frequency could be used as well. This particular processor was chosen keeping in mind the potential code size, the need for a built-in serial port, SRAM for storing variables, and stack, as well as a 16-bit timer for generating interrupts at precise intervals.

The peripheral devices used are Maxim MAX186 ADC, Dallas DS1302 RTC, and Atmel AT24C512 EEPROM. All of these devices have a serial communication protocol. The EEPROM has an IIC interface while the ADC uses SPI link. The RTC uses a simple three-wire serial protocol. Figures 15.9, 15.10, and 15.11 illustrate the signal activity on the AVR processor connected to the three peripheral devices.

The 16-bit Timer1 is programmed to generate interrupts at 10-ms intervals. The Timer1 ISR counts the number of interrupts, and when this count equals the required time, an ADC sample is acquired and the data is written into the EEPROM.

The data is written into the EEPROM such that the MSB is written at a lower address and the LSB is written at the next (higher) address. All entries into the EEPROM are written as a 2-byte word.

The AstroDat also interfaces to two switches which we call "Acquire" and "Menu" keyswitches. The "Menu" key switch is sampled by the program while the "Acquire" key is connected to the INT0 (PD2) signal of the processor. The INT0 is programmed so that a low level on the pin generates an interrupt. The INT0 ISR debounces the key bounce and determines what is to be done with the event. The "Acquire" key has been connected to an interrupt pin so that the program will respond anytime the key is pressed, and there is no need to monitor the key in a polled manner. The "Acquire" key is used to toggle the data acquisition activity. If the "Acquire" key is pressed the first time, it starts the data acquisition process and keeps going till the memory gets full or till the "Acquire" key is pressed again. On the second occasion, the data acquisition process is terminated, and the current data set that is written into the EEPROM is terminated with an end tag corresponding to the start tag. The processor then waits again for either the "Acquire" key or the "Menu" key to be pressed. If the "Acquire" key is pressed, a new data set is initiated by writing a new start tag, and data is acquired and stored in the EEPROM at the programmed rate. If the "Menu" key is pressed instead, the program goes into a user menu mode and communicates to the user though the serial port (a PC terminal is connected to the serial port for this purpose).

When the processor is acquiring data, an LED connected to a PD6 bit called the "Activity" LED is put ON. When the system is waiting for one of the two keys to be pressed, the "Activity" LED is off.

The INT0 interrupt connected to the "Acquire" key is disabled when the system is in user menu mode (after the "Menu" key has been pressed and till the user does not exit the user menu).

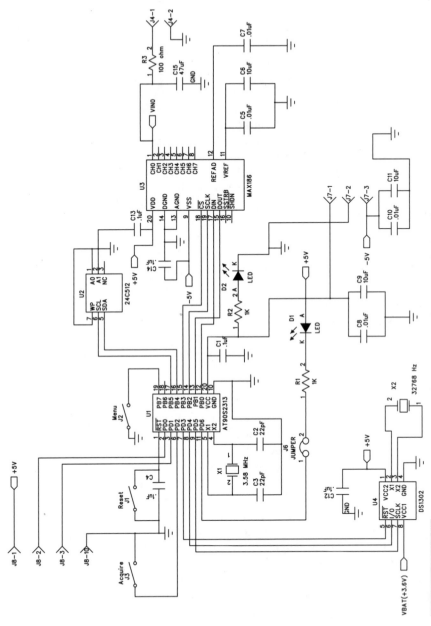

FIGURE 15.8 Circuit schematic for the AstroDat Data Acquisition System.

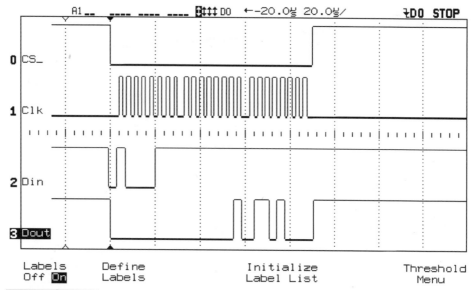

FIGURE 15.9 Signals illustrate the AVR processor controlling the MAX186 ADC.

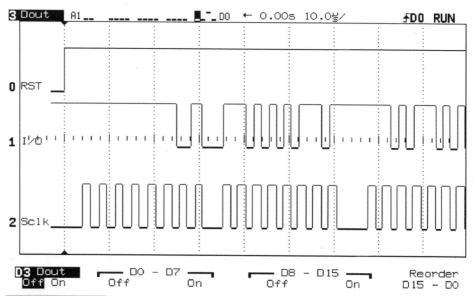

FIGURE 15.10 Signals illustrate the AVR processor controlling the Dallas
DS1302 RTC.

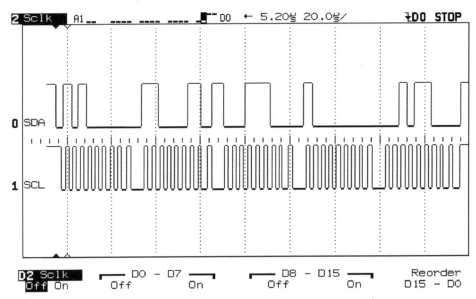

FIGURE 15.11 Signals illustrate the AVR processor controlling the Atmel
AT24C512 EEPROM.

After the user exits the user menu, the user can either press the "Menu" key again or press the "Acquire" key. If the "Acquire" key is pressed, the system starts data acquisition and the "Activity" LED is put ON, and during this period (when the "Activity" LED is ON), the "Menu" key is ignored.

The ADC is an 8-channel, 12-bit sampling ADC from Maxim. We have used the ADC in a single ended, bipolar mode. The ADC has a four-wire serial format link. The processor sends control data on the Din signal of the ADC, and the ADC converts analog voltage on its input, as selected by the control data, and sends the result back to the processor on the Dout signal line. The data is clocked on the serial clock signal SCLK generated by the processor. After the conversion, the chip select signal CS* of the ADC is deactivated to reduce power consumption by the ADC. This signal is activated again by the processor before the start of the next conversion. The signal transitions for the ADC as captured from a logic analyzer are illustrated in Figure 15.9.

The RTC, which is a DS1302 part from Dallas, is operated in a burst mode of operation, and all the date time information is read out. The program tags this information (year, month, date, day, hour, minutes, and seconds) at the start of each data set after the start tag. The user can program the RTC in the "Menu" mode and set the time as well as read the time for verification. The RTC uses an independent 32768-Hz crystal. I have also provided a small 3.6-V Lithium battery for backup power supply. The RTC thus keeps time even if the system is powered off.

Figure 15.10 illustrates the logic analyzer traces for the three signals between the processor and the RTC. All communication starts with the RST signal going high. The control byte is transmitted first by the processor. The data for the control byte is sent on the I/O signal and is clocked into the RTC at the rising edge of the Sclk signal. The data format is LSB first and MSB last. Data from the RTC is clocked out at the falling edge of the Sclk.

15.8   System Development

This system was developed using the proto board described in another chapter. However, even the STK200 evaluation board can be used equally easily. To begin with, small programs dealing with each section of the hardware were developed and tested. Small PCBs were used to build the ADC section, the RTC section, and the EEPROM section, and independently tested. The Timer interrupt was also tested on the proto board for different delay multipliers. Once each of these peripheral devices was tested, they were integrated onto a single PCB and a complete program with all the developed routines was created. It was tested till it started working correctly.

One of the pitfalls in writing to an EEPROM is the write access time. I used the worst-case maximum time of 10 ms as specified in the data sheet. To write more than one byte in the EEPROM, I used the page write mode so that it still takes the write access time of 10 ms maximum to write two bytes. However, if the two bytes are written independently, it will take 20 ms to write, which I avoided by using the page mode of writing.

Eventually the system was connected to a function generator, and multiple data sets were recorded by pressing the "Acquire" key. After the data recording was completed, the system was connected back to the PC serial port, the system was put in Menu mode, and the data was downloaded into a file using the "Write to File" function. Then the data was interpreted manually and plotted on a graph sheet to compare with the waveform generated by the function generator. The comparison was positive, and this completed the system development.

Subsequently, a C program was written to automate the data interpretation and reduction step, and now we can get the plot of the data automatically, as illustrated in Figure 15.12.

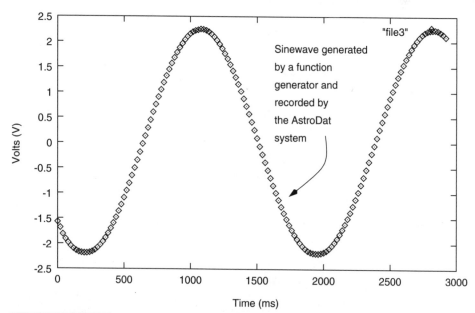

FIGURE 15.12 Sample data plot of a sinewave generated by a function generator and recorded by the AstroDat System. The X axis is time in ms and the Y axis is volts.

15.9   Fabrication

The AstroDat system was prototyped on a general-purpose PCB. All components were socketed for testing ease. The backup battery for the RTC and regulators for the power supply were also placed on the PCB. The system was then connected to an astronomical photometer for field use. The system drew operating voltage (±9 V) from the photometer, which was regulated to ±5 V by the regulators on the AstroDat PCB.

The total current consumed was 12 mA with both the LEDs (power ON and the "Activity" LED) glowing. This power could be reduced further if one decides to put the processor in sleep mode between the Timer1 interrupts (Figure 15.13).

15.10   Design Code

The design code for the AstroDat project is available on the accompanying CD in the code directory in the file astrdat.asm. The code was written as a single program broken up into several subroutines. The code has been written with two interrupt subroutines: one for the Timer1 interrupt and the other for interfacing the "Acquire" key as INT0.

The important design choice that I made early on was regarding the way the data was to be stored. Since the system ADC provides 12-bit data, it would take up two bytes for storing it (although some savings could be done by sharing the unused four bits with another data sample, I decided not to do so). Instead, I used the concept of tagging the data, for ease of data recovery and reduction. Thus the unused four bits are used to tag the data, and looking at the tag one can tell if the word is a data word or not.

Similarly, different tags were used for the time information as well as to mark the start and end of a data set. The format of these tags is illustrated in Figure 15.14.

15.11   Data Readout

Once the data has been acquired in the field and has been stored in the EEPROM, it is eventually brought back to the base. Here the AstroDat system is connected to a PC serial

FIGURE 15.13 Completed AstroDat circuit board inside a plastic enclosure.

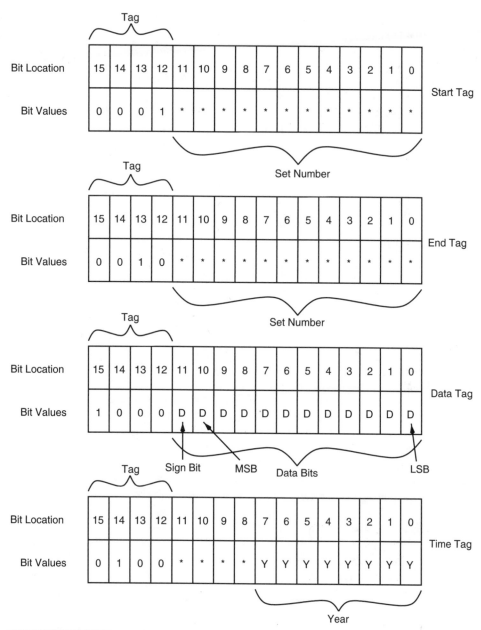

FIGURE 15.14 Format for the various tags.

port. A separate RS-232 to TTL converter PCB was built that connects to the communication port of the AstroDat system at one end and the PC serial port on the other. This was done to minimize power consumption by the RS-232 signal translation components that are not needed during the actual data acquisition process.

The EEPROM data is downloaded and stored into a file as described in the AstroDat User's Guide. The file is an ASCII file and can be inspected. However, I have written a C program to read the file and recover the data for further analysis and plotting. One such plot is illustrated in Figure 15.12.

15.12   AstroDat User's Guide

The AstroDat system operates in two modes: Menu mode and Acquisition mode. The Menu mode is selected when the user presses the "Menu" switch after reset. To operate the system in Menu mode, the AstroDat system is connected to a PC serial port running some sort of terminal emulation program, and the "Menu" key is pressed. The sequence in which the two keys can be pressed is illustrated in Figure 15.15. The data acquisition system designed is used to store data from an analog source with time delay between two acquisitions being programmed by the user.

The device is first initialized by the user at the base. For this purpose, a "Menu" switch is provided. After the system has been initialized as required, it is taken to the field for data acquisition. The data is stored as blocks; each block of data is called a set. The start and end of a set is controlled by the user by pressing the "acquire" switch. The data samples in a set are separated by a time delay, which is programmed by the user at the base. Also, the device has the capability to store the time at which the first observation in a set was recorded.

Data stored in the EEPROM has tags that help to interpret the data. Each set of observations is identified by a unique "set number." Each set of observations will have a start tag (containing the set number) in the beginning of the set and a stop tag indicating the end of the set. The actual data samples are identified by a data tag. Also, each start tag is followed by a time tag that indicates the time at which the first observation of the set was recorded.

System initialization   The "Menu" switch is used for interaction between the system and the user. It helps the user to initialize the real time clock (RTC), change the time interval (delay value) between two observations in a set, download data from the system, etc. To initialize the circuit, the following steps are to be followed:

1) Connect the port P1 to RS-232 board.
2) Connect serial port of the PC to the RS-232 board.
3) Connect power supply to the appropriate pins on the system board.
4) Load VTERM (or any other terminal emulation program) on the computer.

The terminal is set up to communicate with the following features:
  i)   New line—ON
  ii)   Wrap around—ON
  iii) Data bits—8

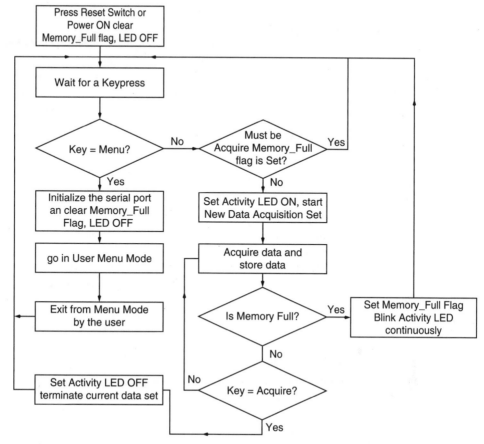

FIGURE 15.15 Flowchart for user interaction using the two keys: Acquire and Menu.

    iv)   Stop bits—1

    v)   Communication rate—9600

    vi)   Parity—OFF

    vii)   Local echo—ON

5) Switch on the power supply to the system board.

6) The Power ON LED should now glow. If it does not glow, either power supply connections are not connected properly, or the LED is faulty.

7) Press the reset switch on the AstroDat system board.

8) Press the menu switch on the board.

9) A menu will appear on the screen with seven choices. Select any one choice by pressing the number corresponding to that choice. Any invalid number will be rejected and a new menu screen will appear.

10) The choices are as follows:

    I)   Initialize RTC

        1) Press 1 to select this choice.

        2) The valid values that can be entered and the format in which to enter them

will be displayed. Data should be entered in the proper format and should have the correct values.

   3) For example, if you wish to enter the following values: 3rd December, 1999, time is 15:01:23 and day is Friday, then enter data as follows:

     a) Enter 23
     b) Press space bar
     c) Enter 01
     d) Press space bar
     e) Enter 15
     f) Press space bar
     g) Enter 03
     h) Press space bar
     i) Enter 12
     j) Press space bar
     k) Enter 05
     l) Press space bar
     m) Enter 99

   4) After the last value is entered, the main menu screen will appear.

Note: Day is chosen such that Monday is day 01 and Sunday is day 07. Time is taken in 24-hour format. To enter 1, the user should enter 01. Single digits are invalid.

II)   Read RTC

   1) Press 2 to select this choice.
   2) The choice will display the current data in the RTC.
   3) The format of each field in the data is displayed on the screen for interpretation.
   4) To go back to the main menu user should press a key.

III)   Read EEPROM

   1) Press 3 to select this choice.
   2) The choice is used to read data stored in the EEPROM, in batches.
   3) At a given time, a certain amount of the data stored in the EEPROM is displayed. To view another batch of data, the user should press a key. Once all the data stored in the EEPROM has been displayed, pressing of a key will display the main menu again.
   4) Displaying the data in batches helps the user to interpret the data as it is displayed on the screen. Do not use this choice to store EEPROM data into a file.
   5) Interpretation of the data stored is given afterwards.

IV)   Enter delay

   1) Press 4 to select this choice.
   2) The choice is used to allow the user to change the value of delay time between two observations in a set. The value of this delay is the same for all sets of data observations.
   3) The delay value is calculated as follows:

     a) Divide the delay value required (in milliseconds) by 10.
     b) Convert the quotient obtained into hex.
     c) Enter the hex value obtained as the delay value.

   4) After the delay value has been entered the main menu screen will appear.

Note: The delay value entered should be of four digits always. The minimum value that can be entered is 0002 and the maximum value that can be entered is FFFF.

V) Read delay
   1) Press 5 to select this choice.
   2) The choice is used to display the current delay value stored in the system. Value is displayed in hex format.
   3. Conversion of the value displayed is as follows:
      a) Convert the value displayed (in hex) to decimal.
      b) Multiply the decimal value by 10.
      c) The product obtained is the actual delay value in milliseconds.
   4) User should press a key to return to the menu.

V) Write to file
   1) Press 6 to select this choice.
   2) The choice is used to display the data stored in the EEPROM without any breakage. After the data is displayed, the user should press a key to return to the main menu.
   3) The choice can be used to store EEPROM data into a file for further processing. Before selecting this choice, appropriate file store/transfer commands should be set up in VTERM.

VII) Exit
   1) Press 7 to select this choice.
   2) The choice is used to exit from menu mode. On selecting this choice, the screen should go blank.

Acquiring data in the field   The "acquire" switch is used to indicate the start and end of a set of observations. Pressing the "acquire" switch when the "activity" LED is off indicates the start of a set. The system now starts taking observations, and this is indicated by a glowing "activity" LED. The time delay between two data acquisitions in a set is programmed by the user, at the base using the "menu" switch. Pressing the "acquire" switch again will indicate the end of the set. The "activity" LED now switches off. The data thus recorded can be read back and analyzed at the base using the "menu" switch.

The following steps should be followed for acquiring data in the field:

1) Connect the power supply to the appropriate pins on the system board.
2) Connect the analog input signal to be stored to the data pin (J4).
3) Switch on the power supply to the system board.
4) The "red" LED should now glow. If it does not glow, either power supply connections are not connected properly, or the LED is faulty.
5) Press the reset switch on the system board.
6) Press the "acquire" switch to indicate the start of a set of observations. The "activity" LED now glows.
7) Press the "acquire" switch whenever the data acquisition for the set has to be stopped. The "activity" LED will now switch off.
8) Press the acquire switch whenever another set of observations has to be recorded.

Note:

1) The power supply, once switched on, should not be switched off even between two sets of data acquisitions, when the system is not actually recording any data. Switching off the power will reset the circuit. When power is restored and the "acquire" switch is pressed again, the previous set of data acquisitions will be over-written. The same precaution should be taken with the "reset" switch. It should be pressed only once after the power supply is given to the circuit.

2) In the field, the "menu" switch should not be pressed at all. This switch should be pressed only when the system has to be initialized at the base. Pressing this switch in the field (when the system is not interfaced with a computer) will hang the system and it will have to be reset.

3) While the data acquisition is in progress, if the "activity" LED starts flickering, then it indicates that the system memory (EEPROM memory) is full and no further samples can be recorded. Pressing the "acquire" switch now will not affect the status of the "activity" LED and the system has to be reset.

Storing of data in the memory The data in the EEPROM (system memory) is stored as illustrated below. The format illustrated can be used for interpreting the data read from the EEPROM.

MEMORY LOCATION	DATA STORED
0000–0001	Time delay value
0002–0003	Start tag + set number
0004–000B	Time tag + time
000C–000D	Data tag + data
.	.
.	.
.	.
(n)–(n + 1)	End tag + set number
(n + 2)–(n + 3)	Start tag + set number
.	.
.	.

Definition of individual tags

1) Start tag + set number
This tag is two bytes. Start tag is 1H (0001)b followed by the set number of 12 bits.

BIT NUMBER	DATA STORED
15	0
14	0
13	0
12	1
11-0	Set number

2) End tag + set number
This tag is also two bytes. The set number in this tag is the same as that in the corresponding start tag. The stop tag is 2H (0010)b.

BIT NUMBER	DATA STORED
15	0
14	0
13	1
12	0
11–0	Set number

3) Time tag + time
The time tag is one byte which is followed by seven bytes of RTC data, which indicates the time at that the first observation in a set was recorded.

BYTE NUMBER	DATA STORED
1	Time tag = 40H
2	Year value
3	Day value
4	Month value
5	Date value
6	Hour value
7	Minute Value
8	Seconds value

4) Data tag + data
This tag is two bytes. The data tag is 8H, which is followed by 12 bits of actual sampled data.

BIT NUMBER	DATA STORED
15	1
14	0
13	0
12	0
11–0	Sampled data value

The sampled data value is interpreted as follows: Bit
11 is the sign. "1" indicates that the data value in
bits 10–0 is in two's complement and the sampled value is
negative. "0" indicates that the data value in bits 10–0 is to be read as it is, and the sampled value is positive.

16

AVR PROJECT 7:
SECURITY DONGLE

16.1   At a Glance

This chapter looks at ways of building security locks, also called dongles, for various applications.

It covers:

1. What security locks are
2. Some ways in which security locks can be designed
3. The algorithm for such locks
4. Building the locks for the serial or the parallel port of the PC
5. Designing a novel security lock for the PC parallel port
6. PC driver for the lock
7. Alternate designs

16.2 Introduction

16.2.1 WHAT ARE SECURITY LOCKS?

What are security locks and what are they used for? Security locks in the context of computer hardware and software refers to a system of authorization that allows use of such hardware or software only to authorized users. As a computer user, you would be aware of login identification and associated password. Unless the correct password for a given login identification is entered, the computer system would not allow you to use the computer. This is an example of a software-based security lock of some sort.

Similarly, when you purchase some expensive software, the software author or the supplier would want some control over the use of the software only to authentic customers and would like to restrict unauthorized proliferation of the software, as this leads to his loss of revenue. To implement this control, the author could provide you with a piece of hardware that is to be connected to the PC system on which the software is supposed to run. When you run the software on a computer, it would look for the presence of this hardware and if it fails to find it, the software would abort and terminate. On the contrary, when it finds the hardware present in the system, the software performs merrily.

Typically, this additional piece of hardware is in the form of a small, sealed printed circuit board with circuitry and with a means of connecting it to the host computer. To allow the user to connect and disconnect this hardware easily, it is usually connected to the host parallel port or the RS-232 serial port, as these are the ports that are easily accessible to the user. Security locks to go on the USB also are available. The lock system works in the following manner:

1. The PC software sends some information to the lock and expects some return information.
2. If the lock has been installed and is performing correctly, it would supply back the expected information, in which case the software continues to execute.
3. If the lock has not been installed or has been removed, the software does not get what it is looking for and so it aborts. Typically, the software periodically queries the lock.

16.2.2 VARIOUS HARDWARE LOCK SCHEMES

There could be many ways in which the software sends information to the lock and expects the return information, and this distinguishes one lock from another. A good lock should not only work as a deterrent against unauthorized proliferation, but should also protect against any hacking attempt. This can be achieved if the lock scheme has infinite combinations (just like a mechanical lock). If a lock scheme always sends one particular type of information and expects a particular return answer, it is no good, as this can be duplicated. Ideally, the lock should have infinite combinations, though in practice that may not be feasible.

A good portable hardware lock system should offer the following features:

1. Should connect to user ports such as the parallel port or the RS-232 port.
2. Should derive power for the lock circuit out of these ports.
3. Should have a large number of combinations so as to resist any duplication attempts.
4. For the parallel-port-based locks, should offer a pass-through parallel port; i.e., the user should be able to connect other peripherals to the parallel port apart from the lock.
5. Application software should periodically detect the presence of the lock by communicating with it.

Figure 16.1 shows a security lock connected on the PC RS-232 serial port. This figure illustrates the circuitry to contain an AT90S2343 processor to indicate that, in principle, the lock could be implemented using this processor. The RS-232 port signals are used not only for communicating with the lock, but are also used to power the lock. The signals on the RS-232 port have +12-V or −12-V voltages. The application software (which employs the lock) should drive these signals (DTR and RTS in the figure) to +12 V, which is then stepped down to a suitable level for the purpose of powering the lock. The other signals illustrated, TxD, and the RxD, are transmit and receive data signals and can be used by the lock to receive and transmit data respectively.

This establishes a basic infrastructure on which a given lock algorithm could be based. Of course, it is assumed that the circuit employed would support the needs of the lock algorithm.

Another way to design the security lock on the PC parallel port is illustrated in Figure 16.2. This scheme is slightly more demanding than the RS-232 port. This is because usually a PC would have only a single parallel port, the use of which cannot be given up (printing, etc.) simply because you want to use it for a hardware lock! To alleviate this possible problem, people have come up with a concept called a pass-through parallel port, which essentially means that the parallel port is a shared resource among many peripherals. In my laboratory, a single parallel port is used to connect to a ZIP drive, an HP printer, and a PADS application software hardware lock, all working in quiet harmony.

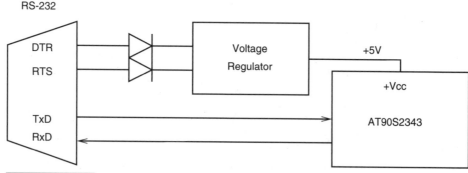

FIGURE 16.1 A security lock on the RS-232 port of the PC.

Parallel Port

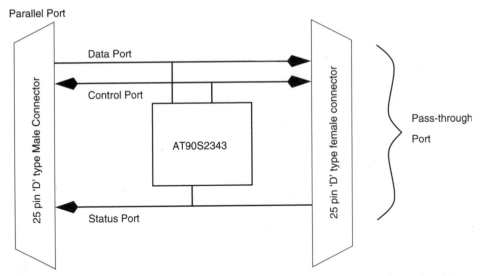

FIGURE 16.2 A security lock on the PC parallel port with a pass-through port.

The ZIP drive is connected first to the parallel port. The hardware lock goes on the pass-through port of the ZIP drive and the printer is attached to the pass-through port of the lock! So this means that the lock circuit in a parallel-port-based lock has to be extra smart compared to that in the RS-232-based lock.

16.3 How to Build an Electronic Lock

Now that we have covered the basics of electronic hardware lock requirements, let us see what it takes to build an electronic hardware lock. We have seen that the lock has a scheme that we call the lock algorithm. Then it must be possible to put that algorithm in the form of realizable hardware. Next, it must connect to one of the user ports for ease of operation. It must derive power from the port. It must use the port for communication with the application software (the software that is supposed to employ the lock) and most of all, the lock must offer pass-through capability.

If we want to build a lock, we must have the lock algorithm. There are many possible lock schemes, from the very elementary to the very complex. An elementary, rather silly scheme (though of equal deterrence) could be a simple wire link from an output signal line of the PC port to an input signal line of the PC port. For the RS-232 port, that would mean shorting the TxD line to the RxD line. Whatever the PC transmits on the TxD line is received back by the PC on the RxD line.

A level of complexity could be built into this scheme by delaying the signal back to the PC by using some kind of memory element. The lock receives the data byte and temporarily stores the byte before transmitting it back on the RxD line. The next level of complexity could be incorporated by employing more memory elements that store a sequence of data bytes received by the lock and then transmitting back in some sequence.

In this chapter, I will show how to build an electronic lock using a scheme that I believe to be quite novel. As I am not aware of the actual lock algorithms that are built into commercial locks, my belief may be more a matter of ignorance than anything else. However, by the end of the chapter you will agree that the scheme I present is indeed an interesting scheme and worth being employed in commercial locks if not already being used.

The backbone of the lock scheme that I am going to present is the Linear Feedback Shift Register (LFSR), which we have discussed in a previous chapter. The lock based on LFSR technique exploits the long repeat cycle feature of the LFSR. An 8-bit maximal length LFSR has 255 (2^8-1) unique combinations. A 20-bit LFSR has a million combinations, and so on.

Figure 16.3 shows an 8-bit LFSR of maximal length. The LFSR is operated by first loading a number (called the preset number or the seed) and then shifting this number. Each shift results in a new number that seems to have no relation to the original number. The 8-bit LFSR could be shifted 255 times before the pattern starts repeating. From the point of view of an electronic lock, a bigger shift register would be very useful. By a hacker, it could be seen as transacting random numbers, frustrating any attempts at breaking the lock.

```
Seed = 1
2   5   b   16  2c  58  b1  63  c7  8f  1e  3d  7a  f4  e8  d0
a1  43  87  f   1f  3f  7f  ff  fe  fc  f9  f2  e4  c8  90  21
42  85  a   14  29  53  a7  4f  9f  3e  7d  fa  f5  ea  d5  aa
55  ab  57  ae  5c  b8  70  e0  c1  83  6   c   18  31  62  c5
8a  15  2b  56  ac  59  b3  66  cc  99  32  65  cb  97  2f  5f
bf  7e  fd  fb  f7  ef  de  bc  79  f3  e6  cd  9b  37  6e  dd
bb  77  ee  dc  b9  72  e5  ca  95  2a  54  a9  52  a5  4a  94
28  51  a2  44  89  12  25  4b  96  2d  5a  b4  68  d1  a3  46
8c  19  33  67  ce  9c  39  73  e7  cf  9e  3c  78  f1  e3  c6
8d  1b  36  6c  d8  b0  61  c2  84  8   11  22  45  8b  17  2e
5d  ba  75  eb  d7  af  5e  bd  7b  f6  ed  db  b7  6f  df  be
7c  f8  f0  e1  c3  86  d   1a  34  69  d3  a6  4d  9a  35  6b
d6  ad  5b  b6  6d  da  b5  6a  d4  a8  50  a0  41  82  4   9
13  27  4e  9d  3b  76  ec  d9  b2  64  c9  92  24  49  93  26
4c  98  30  60  c0  81  3   7   e   1d  3a  74  e9  d2  a4  48
91  23  47  8e  1c  38  71  e2  c4  88  10  20  40  80  1
```

Numbers are in hex.

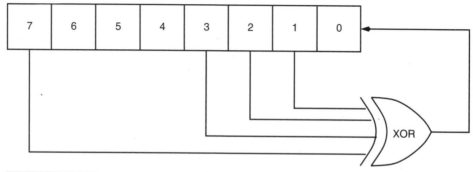

FIGURE 16.3 An 8-bit linear feedback shift register with taps at bit positions 1, 2, 3, and 7.

The 8-bit LFSR can be increased in length to 16, 20, or more bits to provide more combinations in a real situation. For now, let's build a lock based on this simple 8-bit LFSR. This lock, based on an 8-bit LFSR, is proposed to operate as follows:

1. The lock is reset every time it is queried. This assures synchronization between the PC and the processor in the lock.
2. The PC sends two bytes of data. The first byte is the seed.
3. The lock calculates the result and returns it back to the PC.
4. The PC also calculates the result and compares it with the result sent by the lock. When both of them match, the PC concludes that a valid lock is present and then continues executing the application software.

The data transfer between the PC and the lock is serial data transfer with Strobe and Ack handshake signals. The parallel port has three ports, as we have seen in a previous chapter. We use the D0 (DATA port bit0) signal from the parallel port to output serial data from the PC to the lock, the S7 (STATUS port bit7) signal to receive serial data from the lock, the D1 (DATA port bit1) signal from the PC as Strobe to the lock, and S6 (STATUS port bit6) as Ack from the lock to the PC. On the lock side, we use PB0 for serial data input and output, PB1 as Strobe input, and PB2 as Ack output to the PC. Another signal D2 (DATA port bit D2) is used to reset the processor. Figure 16.4 illustrates the block diagram of our scheme. You may note that signal PB0 from the processor is connected to D0 as well as S7 signal pins of the parallel port. These connections cannot be made as it is; we have to isolate the D0 signal pin from the S7 signal pin so that when PB0 is sending data out to S7, the logic level on D0 does not affect the logic levels being set up by PB0.

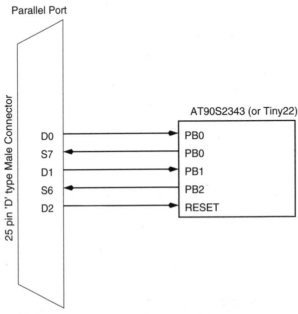

FIGURE 16.4 Block diagram of the lock and the PC parallel port signal configuration.

To achieve this isolation, we use a 1-input AND gate as illustrated in Figure 16.5. The AND gate input is connected to D0 and the output is connected to PB0 as well as S7. Thus when the input of the AND gate is "1", the diode is cut off and the logic level on D0 is isolated from the logic levels being set up by PB0.

Figure 16.6 illustrates the effect of using a passive gate like the 1-input AND gate on logic levels. When the input to the gate is "0", the diode conducts and the output voltage is the input voltage + the forward voltage drop on the diode. If the input "0" voltage is 0.5 V (a valid TTL compatible logic "0") and we use a silicon diode with a forward drop of 0.6 volts, the output voltage would be 1.2 V, which is not a valid TTL level "0". In Figure 16.6, we have used a germanium diode (forward voltage drop 100–300 mV), and the voltage shift on the output signal is about 200 mV, which is a valid logic "0" TTL signal. Figure 16.6 illustrates the hidden dangers of using discrete passive logic and its potential to alter signal levels.

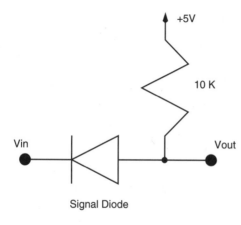

FIGURE 16.5 A 1-input AND gate used as a level isolation circuit.

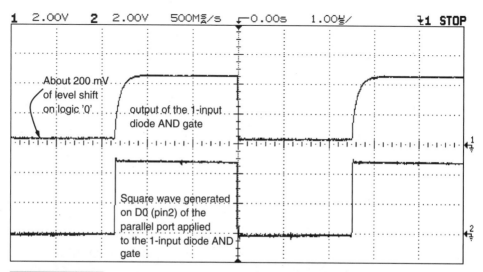

FIGURE 16.6 The effect of driving a 1-input diode AND gate with a logic signal.

16.4   Design Description

Now that we have seen the plan for our version of the electronic lock, it is time to dive into the actual circuit. Figure 16.7 is the circuit schematic for the security lock. The circuit shows a Tiny22 processor, but an AT90S2343 processor with its internal RC clock circuit enabled would do just as well.

Let's plod through the circuit. Connector J1-1 and J1-2 are used to apply power (+5 V) to the circuit. For a real lock, this power would be derived out of the port to which it gets connected. Capacitors C5 and C6 are used to filter the incoming supply voltage. IC U1 is a Tiny22 processor, and I have also tested the circuit with an AT90S2343 processor with its internal RC oscillator enabled. Either of these processor could be used.

Connector J2 is the parallel port DB-25 male connector. This connector mates to the parallel port DB-25 female connector on the PC. Resistor R2 is connected to pin J2-4 (DATA port signal D2), which resets the processor by applying logic "0." Since we have used capacitor C4 on the reset pin of the processor to filter any unwanted noise, R2 limits the discharge current from C4 to damage the port bit D2 of the parallel port. A charged capacitor can discharge large currents (limited by the load resistor), and to limit this current, we have used R2 (470 ohm). The resulting discharge current (about 10 mA, worst case) can easily be handled by the parallel port pin. To reset the processor, the D2 signal is taken to logic "0" for a small time (.001 s) and them taken to "1." This resets the processor.

Pin J2-11 of the parallel port S7 (STATUS port signal bit7) is connected directly to the PB0 bit of the processor and is used to receive serial data transmitted by the processor to the PC.

Pin J2-2 of the parallel port is the D0 (DATA port bit D0) signal pin and is connected to the PB0 pin through diode D1. Diode D1 and resistor R1 form the 1-input diode AND gate that we have already discussed. When the PC wants to receive data from the processor, it first puts signal D0 to "1" and then triggers the processor to send data.

Pin J2-3 is signal D1 (DATA port bit1) of the parallel port and is an output pin from the parallel port used as a Strobe signal by the PC. It is connected to pin PB1 of the processor.

Finally, pin J2-10 is signal S6 (STATUS port bit6) and is an input pin of the parallel port used to receive the Ack signal from pin PB2 of the processor.

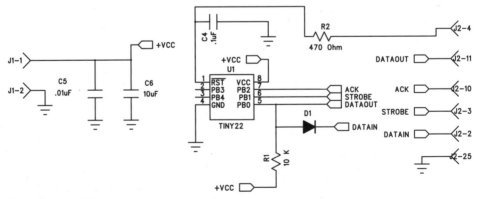

FIGURE 16.7 Circuit schematic for the PC parallel-port-based security lock using AT90S2343.

Let us see how this circuit is able to exchange data. The circuit is set up to transmit data in chunks of 8 bits, 1 bit at a time, i.e., serially. This data transfer between the PC and processor is synchronous, and the Strobe signal acts like a clock when the data is transferred from the PC to the processor, while when the processor transmits data to the PC, the Ack signal acts like the synchronizing clock.

Any data transfer scheme between two devices must ensure that data is always exchanged without any loss. To ensure that, some kind of handshake signals are used. We call these signals Strobe and Acknowledge. Figure 16.8 shows how these signals are used to exchange data between a master device and a slave device. One of the devices is called the master, as it initiates all transfers. The other is called slave, as it always obeys the master. Figure 16.8 shows a data bus, however this scheme is true for exchanging even a single bit of data between the master and the slave, as is our case.

This scheme is also used by the master to receive data from the slave. In this mode, to receive data, the master checks that the master has de-asserted the Ack signal, and after that asserts the Strobe signal. The slave responds by placing the data on the data line and then asserting the Ack signal. The asserted level on the Ack line signals the master that data is available for it to read. The master reads the data and then lowers (de-asserts) the strobe signal. When the slave sees the de-asserted signal level on the Strobe line, it knows that the master has received the data and so it can lower its own Ack signal. This completes one cycle of data transfer between the master and the slave.

I captured the Strobe and the Ack signal activity on a digital oscilloscope, and this is illustrated in Figure 16.9. The Strobe signal is always activated only when the Ack signal is "0" and the Strobe signal goes low after the Ack signal is at logic "1."

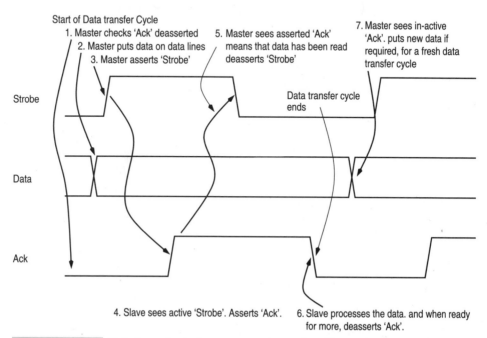

FIGURE 16.8 Data transfer between a master and a slave using strobe and Ack handshake lines.

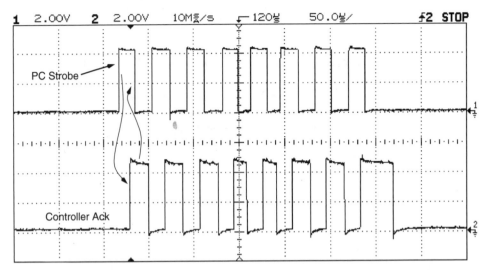

FIGURE 16.9 Scope trace illustrates the time relationship between the Strobe generated by the PC as the master and the Ack by the AT90S2343 as a slave.

Another scope trace is illustrated in Figure 16.10 with data input and output signals (i.e., signals on the PB0 pin) and the PC Strobe. This particular trace was obtained by programming the PC and the processor to transmit a data byte from the PC and immediately receive the transmitted data from the processor.

The trace illustrates the effect of the diode AND gate in the form of a 1.2-V level shift on the logic "0" signal transmitted from the PC. However, this does not seem to be a problem for the processor, as it seems to get the correct data and transmit it back as sent from the latter half of the data trace. This latter half of the data signal is from the processor to the PC (received by the PC on the S7 signal pin J2-11).

What happens if the processor does not deassert the Ack signal? In our scheme the PC must wait till the Ack signal is deasserted by the processor, and this is what exactly happens. I captured the trace of the Strobe and Ack signal between the PC and processor after the PC and the processor code was completed as required for this lock application. Figure 16.11 illustrates the trace. The PC transmits 2 bytes to the processor as we have discussed; the processor then calculates the result, and this could take time, depending upon the value fo the second byte that the PC transmits. During this time the processor holds the Ack signal asserted, thus signaling to the PC that the processor is not yet ready to complete the transaction. The PC waits till the Ack signal is lowered and then proceeds with receiving the result byte from the processor.

16.5   Possible Alternatives

We have used 8-pin processors in this application to reduce size. However, from the point of view of current consumption, that may not be the best choice. AT90S1200A (or the AT90S1200 with its internal 1-MHz RC oscillator) may consume less current than the

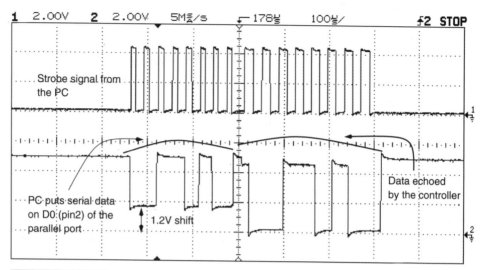

FIGURE 16.10 Scope trace shows 8 bits of data set up by the PC while sending to the processor and the return data generated by the processor.

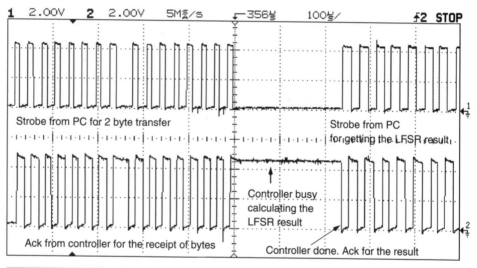

FIGURE 16.11 The scope trace shows how the processor can delay the data transfer back to the PC by asserting the Ack signal. When Ack signal remains "1," the PC waits for it to go "0" before asserting the strobe signal to "1."

Tiny22 or the AT90S2343. This is because that the 1200 device does not have any SRAM and has less EEPROM than either of the two 8-pin devices.

From the point of view of implementation, one could implement a longer LFSR without much impact on the processor code size. Another possible alternative is to put the dongle on the PC serial port. The schematic for just such a design is illustrated in Figure 16.12.

The serial-port-based dongle is implemented using the AT90S2323 with an external 3.58-MHz crystal (nothing special about this value; in fact, any other could be used as well

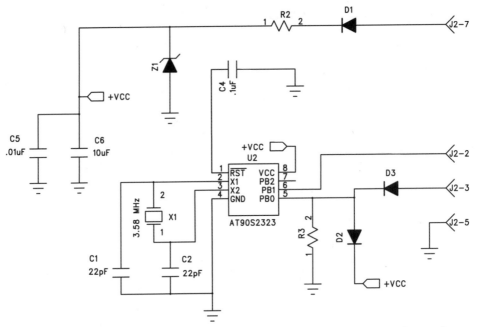

FIGURE 16.12 Circuit schematic for the PC RS-232 serial-port-based security lock using AT90S2323.

with suitable modifications to the software) rather than use an AT90S2343 with the internal RC clock oscillator. This is because the RC oscillator frequency is supply-voltage dependent, and since the power supply is derived out of the serial port pins, which varies, it does not give a constant data transfer rate, which is critical for data transfer on the RS-232 port. To alleviate this problem, the lock was implemented using an AT90S2323 with an external crystal. The serial data transmission from the processor was implemented using the bit banging scheme we discussed in a previous chapter. The data transfer baud rate was selected as 9600 bps, 8 data bits, no parity, and 1 stop bit.

16.6   Fabrication

This little circuit was fabricated on a single-side general-purpose PCB as seen in Figure 16.13. The 25-pin D-type connector was soldered directly onto the PCB.

16.7   Design Code

The code for this project is available in the code directory in the file dongle1.asm for the AVR processor as well as the PC as a file dongle.c. The PC code is in C and was tested on a 100-MHz pentium running DOS with a Turbo C version 2.0 compiler from Borland. The PC driver lacks many features, such as device timeout. As it is, the PC waits indefinitely for the processor to respond. In real application, the PC should wait for some worst-case

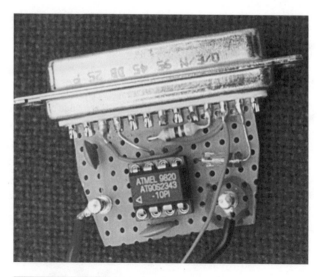

FIGURE 16.13 Photograph of the security
                  dongle.

time and should report "Dongle not found" otherwise. However, the PC code is illustrated
only to highlight the kind of code that will be required to communicate with the processor
through the parallel port.

16.8   Testing

Testing this particular application turned out to be a nightmare for me not because of bad
design or code bugs but because of a faulty power supply. I used this rather expensive +5-
V linear-type power supply to power the lock circuit. However, when I switched the power
supply off, it damaged the processor (which was discovered when the supply was switched
on again and the processor failed to respond to the PC code) due to excessive overvoltage.
Another processor was tried and it met the same fate. I then checked the circuit for any
possible short circuits and, finding none, began suspecting the power supply. Sure enough,
the supply showed 13-V output for a considerable time when it was switched off. This was
captured on the scope and is illustrated in Figure 16.14. Other than this, the system performed as desired.

   The processor results were checked against the expected output of an 8-bit LFSR simulator. The C code that generates the LFSR output is available on the CD in the code directory in the file lfsr.c.

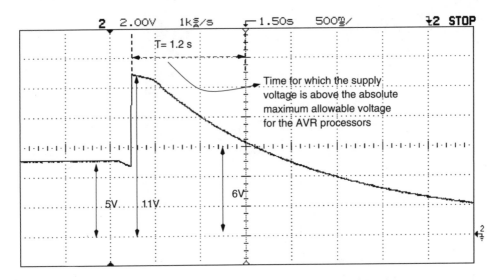

FIGURE 16.14 A case of a bad power supply with the potential to destroy a 5-V-rated processor like the AT90S2343. The trace illustrates the output voltage surging to +12 V when it is switched off.

AVR PROJECT 8: A PULSE FREQUENCY COUNTER WITH AN RS-232 INTERFACE

17.1 At a Glance

In this chapter we look at:

1. What a frequency/period counter is
2. Design of a frequency counter
3. Design of a period counter
4. A working design of an AT90S2323-based frequency counter for the PC RS-232 serial port
5. A practical application for the frequency counter

17.2 Introduction

Measuring frequency can be a frequent requirement. You often need a frequency counter either as a general-purpose test instrument or as a gadget for a specific application. With a general-purpose test instrument, you may want to measure the frequency of a signal, and with a gadget you may want to measure the number of people passing a gate in a minute, for example. Or you may want to measure the wind speed or the speed of rotation of the wheel of your motorbike.

Measuring wind speed or the rotation speed of a wheel may seem unconnected to the matter of measuring frequency, however that is not so. Consider a scheme where you install a magnet of the wheel spokes and a hall-effect sensor on the wheel fork. When the magnet passes the sensor, the sensor output changes state during that time. By measuring the number of such pulses in a unit of time, you calculate the speed of rotation of of the wheel. A similar mechanism could be created for measuring the wind speed using a magnet and a hall-effect sensor on a wind vane.

Similarly, you would also want to measure time between events, which would need a timer. Measuring time and frequency are related issues. Once you can measure time, you can calculate frequency, and vice-versa because time and frequency are inversely related; all you need is some means to process the information.

17.3   How Does a Frequency Counter Work?

Figure 17.1 illustrates the block diagram of a simple frequency counter. This frequency counter is designed to handle analog signals. The analog input is amplified by an amplifier and then passed through a waveshaper circuit to produce digital waves that have the same frequency as the original analog signal. The circuit has a time-base generator that generates precise pulses of required duration. This gate pulse enables a gate (that is why it is called a gate pulse) for the duration of the gate pulse, and the incoming waves are counted by a counter. Just before the incoming pulses are clocked into the counter, the

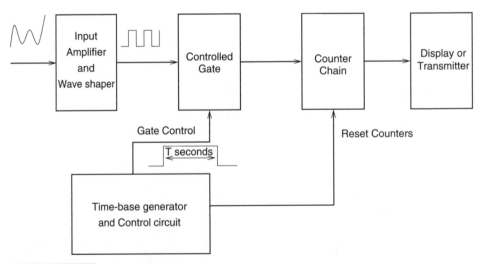

FIGURE 17.1 A frequency counter.

counters are cleared, and at the end of the gate pulse, the contents can be displayed or further manipulated.

If the gate pulse period is set to one second, the count accumulated by the counter is exactly the frequency of the incoming wave in hertz. It is obvious that the frequency to be counted must be greater than the period of the gate pulse. If that is not the case, then either increase the gate period or use the period counter as described in the next section and then calculate the frequency.

Figure 17.2 illustrates the timing diagram of a frequency counter. The time-base signal is passed through a D-type filp-flop to get the gating signal (which is the gate pulse). This results in a symmetrical square wave of 2-Hz frequency with an on period of 1s and an off period of 1s. Other control signals to clear the counters and display the count are not illustrated.

17.4   How Does a Period Counter Work?

The period counter is built in a similar fashion as the frequency counter. Here, instead of the gating pulse gating the incoming wave, the incoming wave is used to gate the pulses generated by the time-base generator. This method is used if the period of the incoming wave is large and measuring frequency is a problem. Consider a situation when the incoming wave has a period of 2 seconds, i.e., a frequency of 0.5 Hz. To resolve the frequency of this wave to some accuracy, we will need to have a time-base gate pulse of say 50 seconds. However, if we choose to measure the period of the pulse, we will get better resolution in a smaller measurement time.

In a period measurement device, the time-base generator generates high-frequency pulses of precise value. These pulses are gated into the counter for one time period of the incoming wave. Figure 17.3 illustrates the block diagram for a period counter. Figure 17.4

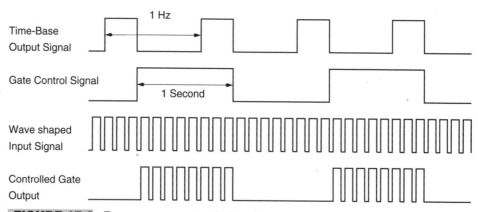

FIGURE 17.2 Frequency counter timing diagram.

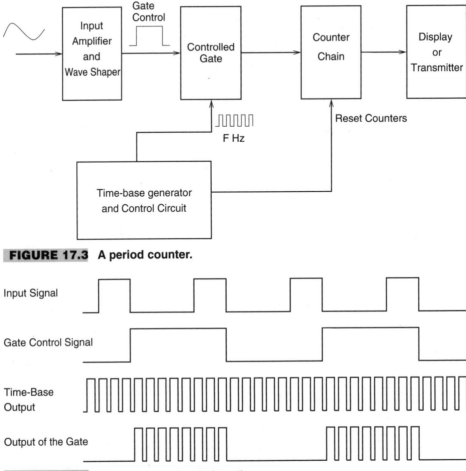

FIGURE 17.3 A period counter.

FIGURE 17.4 A period counter timing diagram.

is the associated timing diagram.

As illustrated from the block diagrams for the frequency counter as well as the period counter, such a device would need many ICs for implementing the various functions. Such a device could be easily implemented using a microcontroller for reducing the component count as well as for providing many additional functions of data manipulation.

An AVR processor is capable of implementing a dual period/frequency counter function with a minimum of components, as illustrated in Figure 17.5. The AVR ports are capable of driving LED displays directly; also, internal timers could be used for the time-base generation and counting functions. All that would be required would be an external amplifier that would provide digital, TTL compatible signal to the AVR. The figure illustrates some switches connected to the AVR for selcting time-base frequency or mode, etc. Besides, the built-in serial port of the AVR provides additional connectivity to a PC for remote control of the instrument or for downloading data for further manipulation or analysis.

The next section discusses the design of a very compact AVR-processor-based frequency counter. The design is expandable to include more features as desired.

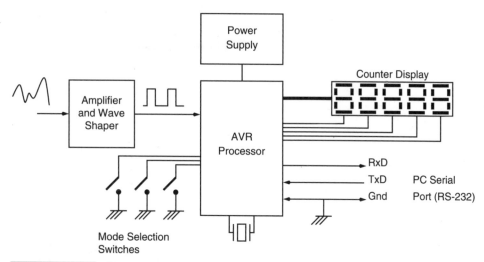

FIGURE 17.5 **A compact multifunction period/frequency counter.**

17.5   Design Description of an AVR-Processor-Based Frequency Counter

This section discusses an AT90S2323-processor-based frequency counter with selectable gate period. The selected processor has just enough I/O lines to permit use as a frequency counter; in fact, all the three I/O lines of the processors are used in this design.

Figure 17.6 illustrates the block diagram of the frequency counter. The features of this frequency counter are:

1. Accepts TTL-level digital signals whose frequency is to be measured.
2. User interface is provided through a PC RS-232 serial port.
3. A choice of three gate pulse periods: 0.1s, 1s, and 10s.
4. Does not require an external power supply. The circuit derives the required power from the RS-232 port of the PC.
5. Uses only a handful of components.

The objective for this design was to build a frequency counter that was very small in size and could count the frequency of digital signals of frequency up to 10 KHz and with different gating periods as listed above.

Another objective was to avoid using an external power supply, thus the choice to use the RS-232 port to draw power was a good choice. However, it also meant that the circuit should be low power and should manage in a few milliamps of current, which is usually available from an RS-232 port. From the large selection of the AVR processors, many processors could meet this design objective. Ideally, I would have liked to use an AT90S2343 and use the internal 1-MHz RC oscillator so as to minimize component count. However, it was found that the internal RC oscillator frequency has a large dependence on the supply voltage, and since the supply voltage for the project was to be derived from the

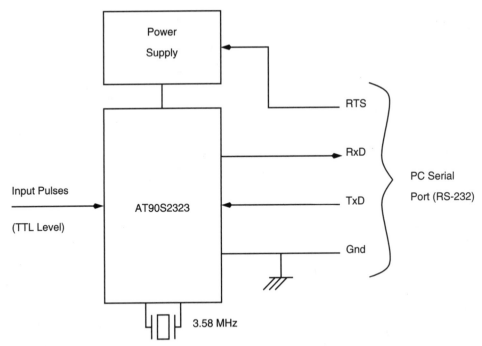

FIGURE 17.6 An AVR-based frequency counter with an RS-232 interface.

RS-232 port and hence expected to be not so stable, this did not seem a feasible processor to use. I then decided to use the AT90S2323, which is very much like the AT90S2343 except that it requires an external crystal. The AT90S2323 has 128 bytes of internal SRAM but no built-in UART (serial port). So it was decided to create a software-driven serial port.

Figure 17.7 illustrates the circuit schematic for the frequency counter. The power to the circuit is derived out of the RTS signal of the PC RS-232 port. Diode D1, resistor R2, and zener Z1 generate the required supply voltage. The diode is 1N4148 signal diode and is used to ensure that only positive voltage is applied to the circuit. Zener Z1 is selected to be 5.1 V, and R2 is 470 ohm to limit the current into the zener diode. Capacitors C5 and C6 are used as supply filter capacitors.

The circuit is operated with a 3.58-MHz crystal. Any other crystal could also be used. In fact, a smaller-value crystal would lead to reduced current consumption by the circuit, however it would also restrict the range of input signal frequency that can be measured by the frequency counter as well as the minimum pulse width of the signal frequency.

Pin PB0 of the processor is connected to the TxD signal pin of the RS-232 port (Figure 17.8). Pin PB0 is programmed as an input pin. Diodes D2, D3, and resistor R3 are used to clamp the positive swing of the TxD signal to within the supply voltage of the processor. When the TxD signal is -ve, the diode D3 blocks it and the resistor offers a logic low to the PB0 pin. The effect of D2, D3, and R3 in restricting the incoming bipolar RS-232 signal to a clamped and rectified TTL signal is illustrated in the oscilloscope trace in Figure 17.9.

Pin PB1 of the processor is programmed as an output pin, and this pin drives the RxD signal pin of the RS-232 port. Please note here that the processor is not generating legal RS-232 voltage swings. However, 0 volts to an RS-232 input is taken as a marking signal and I found that the circuit worked without any problems on a variety of PC machines. The PB1 pin swings between 0 volts on one end and the supply voltage on the other.

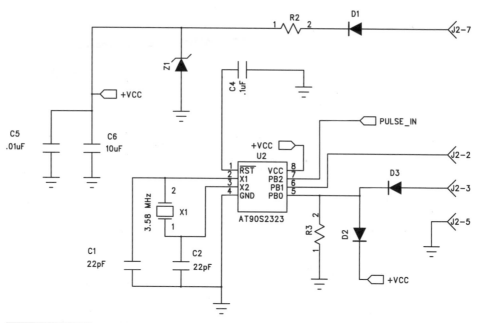

FIGURE 17.7 Circuit schematic for the frequency counter with an RS-232 interface.

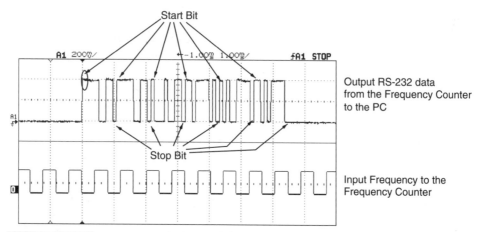

FIGURE 17.8 Logic analyzer trace of the data transmitted by the frequency counter to the PC and the input frequency to the frequency counter.

Pin PB2 of the processor is programmed as an input pin, and the external signal whose frequency is to be measured is applied to this pin. The Timer0 is used for two purposes in this project. When the processor needs to communicate with the PC, the Timer0 is used to generate the time ticks for the serial data transmission bit times. When the frequency counter needs to measure the frequency of the incoming signal, Timer0 is used to generate the gate period.

To begin a measurement cycle, the Timer0 is set up to increment either using the system clock CK frequency (which is 3.58 MHz) in the .1-s and 1-s gating period case, or CK/8 when using the 10-s gating period mode.

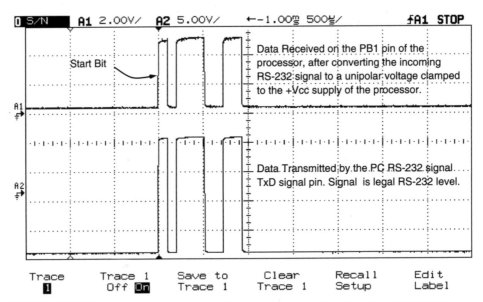

FIGURE 17.9 Logic analyzer trace of the data transmitted by the PC RS-232 port on the TxD pin and rectified and clamped to convert to a unipolar, TTL-level signal on the PB1 pin of the frequency counter.

After the measurement cycle is over, the frequency counter transmits the frequency of the incoming wave to the PC, and for this, it programs the timer to generate ticks for the bit interval times. The frequency counter communicates with the PC at 9600 bits/s, 8 data bits, no parity, and 1 stop bit (9600, 8, N, 1).

The software-driven UART inverts (or complements) the incoming RS-232 signal to account for the signal inversion on an RS-232 line, and before transmitting any value on the line, it complements the number and then transmits the bits.

17.6   Usage

This frequency counter was designed specifically for interfacing to a pulse output astronomical photometer. However, it can be used easily with any sensor that has a frequency output.

17.7   Fabrication

This small circuit was fabricated on a general-purpose PCB measuring 3 cm by 3 cm. A 9-pin D type, female connector was mounted on the PCB for mating to an RS-232 cable.

17.8   Design Code

The design for the project is available in the code directory in the file avpulse.asm. The design code is split up in a few subroutines. These subroutines are as follows:

1. Reset Initialization: initializes the pins, the timer, the stack pointer, and other variables.
2. Main Program Loop: Prints a welcome message on the PC screen and waits for a user selection to select one of the three gate periods, .1 s, 1 s, and 10 s. It then initializes the timer accordingly aad goes on to count the number of pulses for the duration of the selected time. The accumulated count is converted to BCD, the timer is reinitialized to provide serial port bit time intervals, and the result is displayed on the screen. The timer is then reinitialized to provide the gate period, and so on.
3. Get Sample Number: This gets the user response for the gate period value.
4. Collect Data: Counts the pulses on the signal input pin and stores the count.
5. Print Data: Prints the BCD value of the count on the serial port output pin.
6. Get Byte: Receives a byte from the PC serial port at 9600, 8, N, 1.
7. Send Byte: Sends out a byte to the serial port at 9600, 8, N, 1.
8. Tx Msg: Transmits the initial welcome message.
9. Timer0 ISR: This is the Timer0 Interrupt Subroutine that occurs during the pulse count measurement cycle. The ISR determines if the gate period is complete or not.

17.9 Testing

The frequency counter was tested with the help of a Wavetek signal generator, and the readings generated by the frequency counter were compared with the readings on an HP oscilloscope (which has a built-in frequency readout mode) (see Figure 17.10). The results are plotted for the 1-s gate period and are illustrated in Figure 17.11 and also

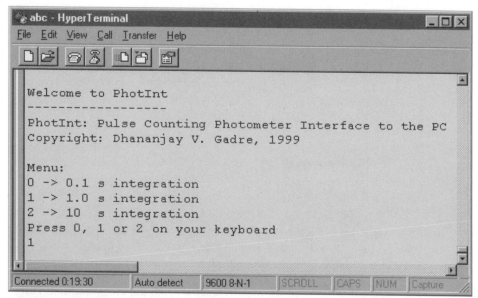

FIGURE 17.10 User interface for the frequency counter.

listed below.

FREQUENCY INPUT (HZ)	MEASURE FREQUENCY (HZ)
50	50
100	101
200	200
500	501
1000	1004
1200	1201
2000	2002
3000	3001
5000	4998
8000	7999
10000	9995
12000	11998
15000	15000
18000	17997
20000	19996

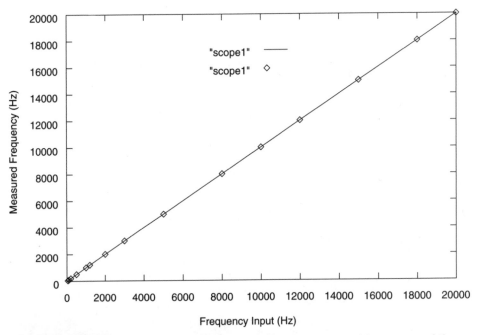

FIGURE 17.11 Plot of the input frequency and the measured frequency of the frequency counter.

18

AVR PROJECT 9: SA-RE-GA

FOLLOW ME—A MUSICAL TOY

18.1   At a Glance

In this chapter, we build an interesting musical memory game. The features of this chapter are:

1. Describe the battery-operated musical memory game.
2. Explain the design based on the AT90S2313 processor.
3. Show how musical notes can be generated.

18.2   Introduction

This is a simple musical memory game. The toy has a set of four switches and four LEDs. It has a small piezo speaker that generates musical notes. Press a switch and a note is produced and an LED glows for the duration of the note. Press another switch and another note is produced and another LED glows. Each switch and LED is associated with a unique note. To begin with, the toy produces a random note when you press any switch, and the LED associated with that note also glows. Then you regenerate that note by pressing the right switch. If you guessed right, the game proceeds to the next level and produces two notes,

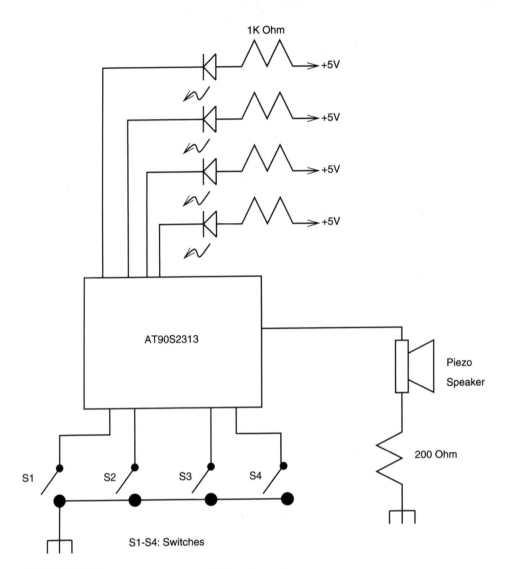

FIGURE 18.1 Block diagram of this musical toy.

retaining the first note, followed by another note selected at random. You then generate these notes in the correct order and so on. If you fail, you can start again. If you succeed, you go to the next level. When you reach the last level, you have won and the toy hails you with a congratulatory note sequence.

This game was designed for my son but I end up playing it more. I have been told that this game is a good test for your musical abilities. If you can remember and regenerate a long sequence of random, uncorrelated notes, you have a musical virtuoso inside you. Figure 18.1 illustrates the block diagram of the musical toy. The processor is operated at 4 MHz using an external crystal.

18.3 Design Description

Let us now consider the design of the toy. The processor chosen for this project is AT90S2313, considering the amount of I/O (total 9 I/O pins required) and the software complexity. An AT90S1200 was considered, and considering the software complexity, I decided to use 2313 for this project, the reason being the need for SRAM for storing the random notes that would be played during the course of the game. These notes need to be stored for later comparison with the user response. Since the AT90S1200 is not equipped with any SRAM, it could not be employed.

The Timer0 is employed to generate the notes. The timer interrupt is used to occur at twice the rate of the required note frequency. The Timer ISR then toggles the output bit, which is connected to the piezo buzzer, generating a note at the required frequency (see Figure 18.1).

The selected notes were of the frequencies 440 Hz, 494 Hz, 523 Hz, and 587 Hz. Timer reload values corresponding to these notes were calculated for a clock frequency of 4 MHz.

At reset and each time the user loses and starts to play again, the program creates a table of 32 entries with random numbers using the Linear Feedback Shift Register (LFSR) principle. The Timer0, which is free-running, is used once at this time to get a seed number for the LFSR algorithm. After this the Timer0 is used only for generating audio notes. The program then waits for a key, any key, to be pressed and plays the first note using the random number table entry. The LED corresponding to this note is also lit up. It then waits for the user to press a key. If the key matches the note, the note is played again, the LED is lit up again, and the program proceeds to the next level. Now it generates the first note again and another note using the second entry from the random number table. Again, after playing the notes, the program waits for the keys to be pressed in the right sequence. If the keys match the note played, the program proceeds to the next level. This can go on until all the 32 notes have been played. If the user has been able to play back the correct sequence for all the 32 notes, the user wins the game. Else, the program starts again, creating a new random number table. Before calling the random number generator, the interrupts are disabled. The routine to generate the random numbers is a critical section of the code that needs to run without being interrupted. After returning from the random number routine, the interrupts are enabled again.

Figure 18.2 illustrates the circuit schematic for the toy. The switches are connected to the PORTD pins and the LEDs and the speaker to the PORTB pins. This arrangement was dictated by Atmel's AVR evaluation board on which the prototype was tested. The switches do not have any pull-up resistors as the internal pull resistors in the PORT pins are activated. The LEDs are connected to sink current into the processor pin.

18.4 Fabrication

This toy was initially developed on the Atmel's AVR evaluation board. After that, it was built on a general-purpose PCB. It runs off a 9-V battery and a 78L05 regulator. Instead of a 9-V battery, even four 1.5-V cells could be used.

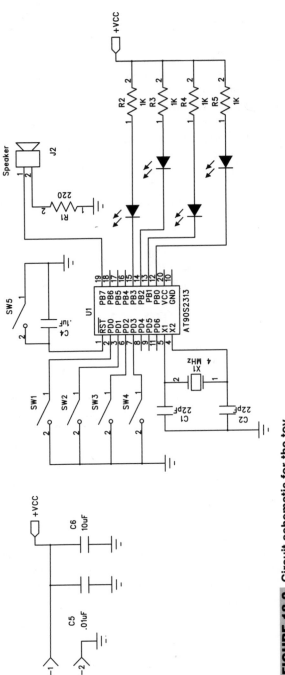

FIGURE 18.2 Circuit schematic for the toy.

18.5 Design Code

The design for the project is available in the code directory in the file toy3.asm. The code for the system was developed and tested in small pieces to begin with. Some of the subroutines were taken from earlier-developed code, e.g., the LSFR-based random number subroutine was taken from the Dongle project. Similarly, the audio tone generation was modified from the Morse keyer project. Here, of course, the tone generation sunroutine in the form of the Timer0 ISR was a little different, as it had to generate four different tones. In fact, the Timer0 ISR is the most critical section of the code, and let us understand how it works.

Even though the AT90S2313 has SRAM that is used for stack, the ISR uses the save_status register (register R0) to save the machine status register SREG. This takes fewer clock cycles than pushing the SREG register on the stack. However, please note that if your application uses the LPM instruction, then R0 register should not be used to save the machine status in an ISR, as the LPM instruction uses the R0 register. Instead, any other register (say R1) could be used.

The ISR then puts off all the LEDs and checks if the audio note is to be played or not. The note is played for a short time only when the status flag is set up. After the time for playing the note is over, the ISR clears the status flag and no note is played till the main program sets up this flag again.

Depending upon the note to be played, the ISR loads the TCNT0 register accordingly. It also lights up the LED corresponding to the note being played. In the end, it toggles the output bit (PB7) which is connected to the piezo buzzer.

If the note-playing time is over, all the LEDs are put off and the output audio bit (PB7) is set to "0." Before the ISR returns to the interrupted program, the SREG processor status register is popped back from the R0 register.

```
;This code segment code cannot run as it is, it is here as an
;illustration
Timer0_int:
        in save_status, SREG
        sbi PORTB, 0     ;first put off all LEDs
        sbi PORTB, 1     ;on PORTB
        sbi PORTB, 2
        sbi PORTB, 3
        sbis PORTB, 7    ;this code increments
        inc count        ;count in alternate ISRs
        cpi count, PLAY_TIME
        brne still_time  ;if play duration is over
        ldi stat_flag, 0 ;reset the flag
still_time:              ;see which note to
        cpi play_t, 0    ;play
        brne chk2
        ldi play_t, T1
        out TCNT0, play_t
        ldi play_t, 0
        sbrc stat_flag, 7 ;first check if note is
                ;being actually played
                ;if so,
        cbi PORTB, 0     ;put on LED on PORTB0
        rjmp chk5
```

```
chk2:    cpi play_t, 1
         brne chk3
         ldi play_t, T2
         out TCNT0, play_t
         ldi play_t, 1
         sbrc stat_flag, 7
         cbi PORTB, 1      ;put on LED on PORTB1
         rjmp chk5
chk3:    cpi play_t, 2
         brne chk4
         ldi play_t, T3
         out TCNT0, play_t
         ldi play_t, 2
         sbrc stat_flag, 7
         cbi PORTB, 2      ;put on LED on PORTB2
         rjmp chk5
chk4:    ldi play_t, T4
         out TCNT0, play_t
         ldi play_t, 3
         sbrc stat_flag, 7
         cbi PORTB, 3      ;put on LED on PORTB3
                           ;check if flag to play note
                           ;is set to 255
chk5:    cpi  stat_flag, 255
         breq play_it
         cbi PORTB, 7      ;if no, then clear PB7, so
                           ;that the speaker does not load
                           ;PB7
         rjmp no_tone      ;note is not to be played
                           ;so just return back
                           ;if flag is set, then play note
                           ;for that just complement PB7
                           ;stat_flag is being used as a temp
                           ;register. Its value is restored
                           ;later.
play_it: in stat_flag, PORTB
         ldi temp2, $80
         eor stat_flag, temp2
         out PORTB, stat_flag
         ldi stat_flag, 255
no_tone: out SREG, save_status
         reti
```

Figure 18.3 illustrates one of the notes being played by the toy. Now that the toy is working well, my son and I are having a great time playing with it.

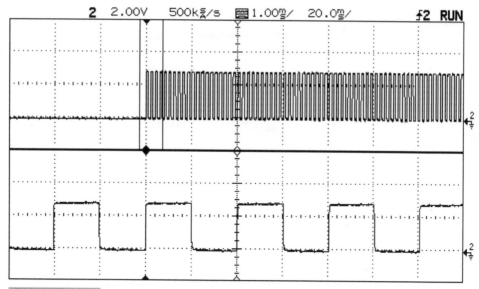

FIGURE 18.3 A digital oscilloscope trace of the tone generated by the toy.

AVR PROJECT 10:
AVR PROTOBOARD™ FOR NUTS™

19.1   At a Glance

1. A simple and inexpensive application prototyping board.
2. Uses the STK200 dongle for program download.
3. Uses Atmel AVR ISP software for project management.
4. Offers a general-purpose prototyping area to add custom hardware.
5. Allows all classic 20-pin AVR controllers to be used.
6. Can be modified to suit other AVR processors.

19.2   Introduction

This chapter looks at a simple and inexpensive prototyping board for developing AVR applications.

However, I want to make it clear that some really inexpensive and compact prototyping and evaluation boards are available through Atmel. These boards have more features and hardware capability than the one described here, and I strongly recommend that users buy one or more of those boards (STK200, STK300 etc.). These boards are priced very competitively.

Therefore, only such users who want to quickly evaluate AVR processors before seriously committing any resources or those who are really constrained should consider building the protoboard described here.

This project is targeted towards Nuts (no offense intended), elsewhere called Dummies. Here is my own, very own, Protoboard for these Nuts.

The design is suitable for use in a college laboratory (which has access to a PC) or a NutShack. The design allows an experimenter Nut (henceforth called eNut) to play with all the hardware features of the AVR controller: serial port, timers, analog comparator, interrupts, and various sleep and power-down modes.

The in-system programming circuit allows the eNut to download programs from the PC parallel port to the protoboard.

Figure 19.1 illustrates the block diagram of the protoboard and Figure 19.2 illustrates how the protoboard connects to the PC development system through the parallel-port-based dongle. The AVR protoboard has an ISP port, and the parallel port dongle connects to this port through a ribbon cable and a mating box connector.

19.3   Design Description

The prototype board is not much of a design. The circuit consists of a crystal oscillator, which I have set to 3.58 MHz, a MAX232 chip for RS-232 data translation, and an ISP programming header for serial programming of the AVR chips. A manual reset circuit is

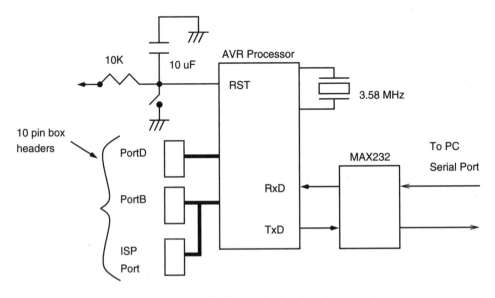

AVR Processor Experimentor's Board

FIGURE 19.1 AVR Protoboard for Nuts.

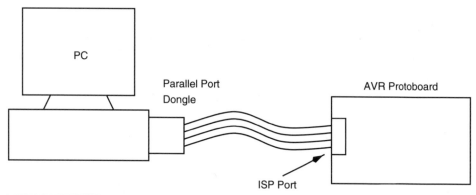

FIGURE 19.2 Connecting the AVR Protoboard to the PC for program download using the ISP port.

also provided. Figure 19.3 illustrates the prototype board circuit and Figure 19.4 is a photograph of the completed circuit board. The PortD and PortB signals are terminated on 10-pin box-header connectors. Users can create their own circuits with matching connectors for connecting peripheral devices. The circuit must be operated at a regulated +5-V power supply.

The programmer dongle that connects to the PC through the PC parallel port on one end and the ISP header on the header is illustrated in Figure 19.5. The circuit consists of a 74LS244 buffer, a capacitor, a signal diode 1N4148, and a 100-K resistor. The '244 buffer circuit ensures that the parallel port signals can drive the serial programming signals of the AVR processor in the target circuit of the protoboard. The buffer is used to write and read program memory and EEPROM (as desired) to and from the AVR processor into the PC. Figure 19.6 is a photograph of the dongle circuit board.

This dongle and the prototype board use the Atmel AVR ISP software developed by Kanda Systems, and they have very kindly provided a copy of this software to be included on the accompanying CD.

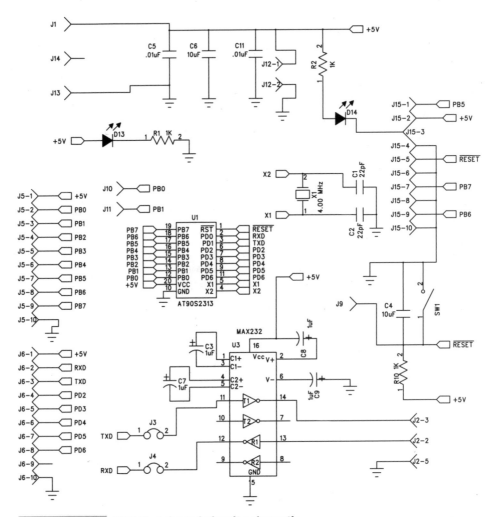

FIGURE 19.3 AVR Protoboard circuit schematic.

FIGURE 19.4   Photograph of the completed AVR Protoboard.

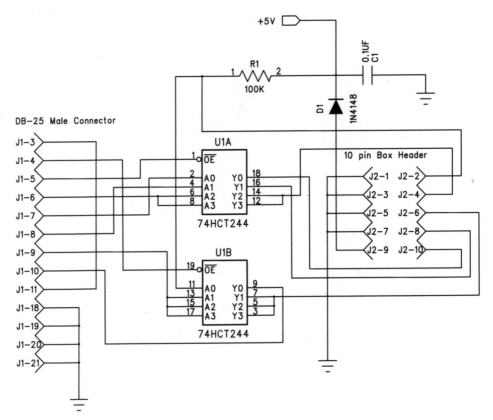

FIGURE 19.5   Printer port dongle to program the AVR Protoboard.

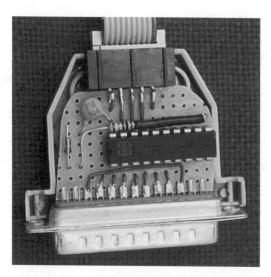

FIGURE 19.6 The printer port dongle to connect the AVR Protoboard to the PC.

IDEAS FOR PROJECTS

20.1   AT90S2343 Controller-based Code Authenticator

The idea of this circuit is to provide a very simple code authentication scheme. The circuit is based on the AT90S2343 AVR controller. It has three input keys and two output LEDs for visual feedback. These LED outputs are also used to provide a "Valid" signal when the entered key sequence is correct.

The input keys are labeled "0," "1," and "New." To enter a key sequence, the "New" key is pressed followed by the sequence of keys for the code, "0" or "1." After each key press, the "Ack" LED indicates that the key press is recognized. For an error, the "Ack" and the "OK" LEDs blink alternately. If the entered key sequence is correct, both of these LEDs go ON at the same time and stay on for five seconds (Figure 20.1).

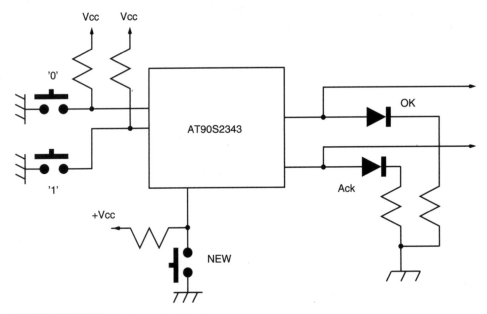

FIGURE 20.1 Code authenticator.

20.2 A CCD Camera Controller

CCD cameras have become extremely popular. You can use the AVR controller to build a CCD camera yourself. Figure 20.2 illustrates the block diagram for a CCD camera, and Figure 20.3 illustrates how the CCD Camera connects to the PC. You could choose one of the many communication links we outlined in a previous chapter. Some popular CCD chip manufacturers include Texas Instruments and Kodak.

20.3 Personal Temperature Logger

This little project offers you the ability to record your body temperature using a minimal possible system using just two 8-pin ICs, a controller, and a temperature sensor! The controller could be the tiny 8-pin AT90S2343. The temperature sensor is LM75 from National Semiconductor. The circuit is so small that it can be worn as a locket around the neck, and the temperature probe can be stuck somewhere on the body using an unmedicated Band-Aid sticker. Neat, isn't it? (Figure 20.4)

The temperature logger features:

1. 0.5 degree C temperature resolution.
2. Temperature range: −25 C to 100 C.
3. 128 data points nonvolatile storage.

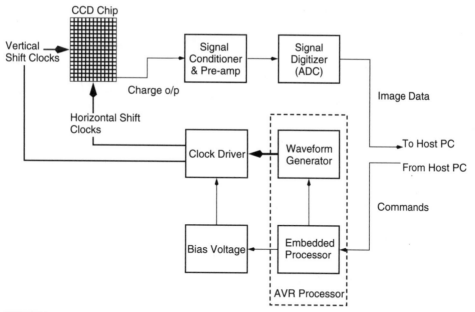

FIGURE 20.2 Block diagram of a CCD camera controller.

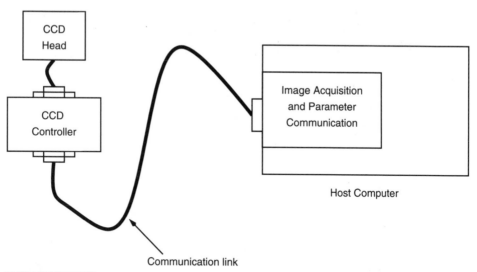

FIGURE 20.3 CCD camera connectivity to the PC.

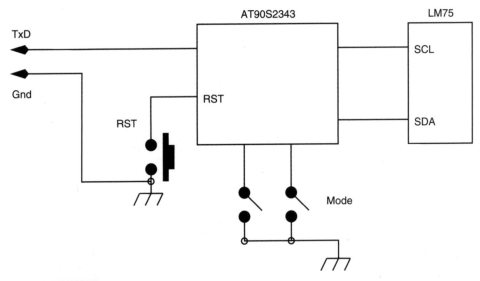

FIGURE 20.4 A personal temperature logger.

4. Selectable record rates: 15 min, 1 hr, and 3 hr.
5. Estimated active current consumption (during sampling): 2 mA. During sleep mode, this falls to less than 1 mA.
6. Operational voltage: 3 V (2 × 1.5-V cells).

20.3.1 CONFIGURING THE TEMPERATURE LOGGER

The two mode switches have four possible settings: speed1, speed2, speed3, or readout. Set the two mode switches to select the appropriate record speed (speed1, speed2, or speed3), press the reset button and let go. The circuit starts logging temperature.

20.3.2 Extracting Data

To read out the data, set the two mode switches to readout mode (both switches OFF), connect the logger to the PC serial port, and press the reset switch. The data logger dumps the data to the PC at 2400 baud, 8N1. A suitable terminal emulator program can save the data to a file for analysis or viewing.

20.4   Swipe Card Reader

A Swipe card contains bar-coded information as seen on many products. I propose to use an IR LED and detector to scan a bar-coded swipe card using AT90S1200 controller. The controller reads the swipe card and dumps the code onto a LCD display, a serial port, or it could be used to activate some valve/relay, etc. The swipe card idea is illustrated in Figure 20.5.

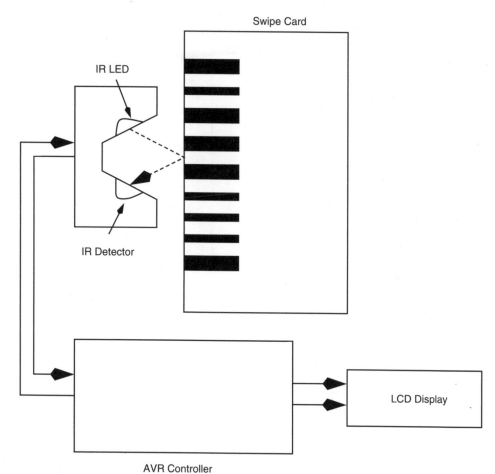

FIGURE 20.5 A swipe card reader.

20.5   IBM PC Keyboard Decoder

It is sometimes desirable to interface a PC/AT keyboard to a microcontroller for some applications. This project illustrates how to achieve just that. The diagram in Figure 20.6 illustrates the idea. The keyboard is connected to the AT90S2313 controller through connector J1. Only two lines are required: for the clock and data signals of the keyboard. These lines are bidirectional. During normal operation, the keyboard drives the clock and data lines; however, the controller can also take control of these lines to send commands to the keyboard.

20.6   A Morse Code Tutor

Connect an LCD display to an AVR controller, add a few switches and a speaker and you are ready to build this Morse code tutor illustrated in Figure 20.7. A partially completed circuit board is illustrated in Figure 20.8.

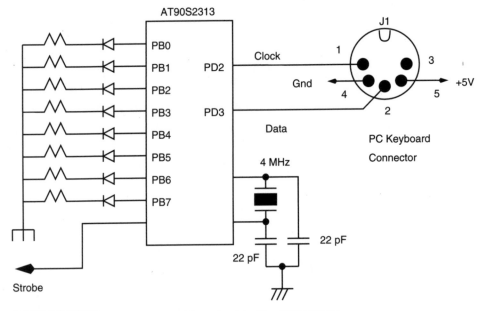

FIGURE 20.6 A PC keyboard interface to the AT90S2313.

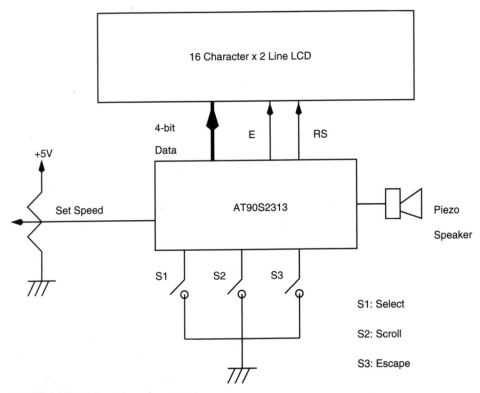

FIGURE 20.7 A Morse code tutor.

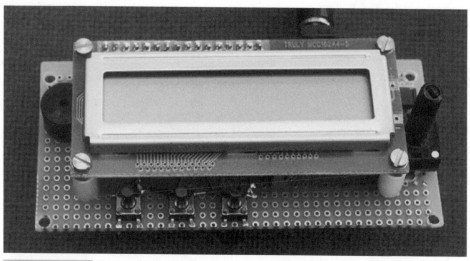

FIGURE 20.8 Photograph of the Morse tutor circuit board.

ADC: Acronym for Analog to Digital Converter. An electronic device or system that encodes analog voltage/current to a multilevel digital number.

address bus: A group of signals in a microprocessor system that indicates the address of the memory location from where the data is to be read or written to.

ASCII: American Standards Committee for Information Interchange. A 7-bit coding scheme for alphabets, numerals, punctuation, as well as control characters.

ASIC: Acronym for Application Specific Integrated Circuit.

assembler: A software program that takes a text file called the source file, and converts it into another file with the machine op-codes (simply called the machine code or object code).

BCD: Acronym for Binary Coded Decimal.

BOD: Brownout Detector. An electronic device that detects a drop in supply voltage below a threshold and generates a signal to reset the processor till the supply voltage is restored to acceptable level.

BIT: Binary digit.

bps: Acronym for Bits per Second.

byte: A number with a maximum of eight bits. Thus the byte-wide number is between 0 and 255 (decimal).

CISC: Acronym for Complex Instruction Set Computer. This a type of processor architecture that is characterized by a variable instruction length, usually small numbers of register and multiple address modes.

compiler: A software program that converts a high-level language source code into the machine language that the processor can execute.

counter: A register that is incremented for each occurrence of an event on an input pin of the counter. The event is indicated by a pulse. For each pulse, the counter is incremented by one.

critical section: A chunk of code that must be executed without any interruption for proper operation.

CPLD: Acronym for Complex Programmable Logic Device. A programmable logic device rich in gates and interconnection circuitry.

CPU: Acronym for Central Processing Unit. The CPU is the computational and control unit of a computer.

DAC: Acronym for Digital to Analog Converter. A device or a process that converts a digital number to a corresponding analog voltage or current.

data bus: A set of signals that carries the data information between the processor and the memory and/or I/O devices.

debug: To correct mistakes in a program.

debugger: A software program that assists in debugging a piece of code.

development host: A computer system that hosts development software like an assembler, compiler, programmer, debugger, etc.

dongle: A mechanism to ensure that only authorized users can use a particular software with the help of an electronic lock provided with the software to deter software piracy. The dongle needs to be connected to the PC parallel port, serial port, or the USB to be able to use the software.

duplex: Term used in communication systems that means that both the transmitter and the receiver can send and receive signals at the same time.

embedded controller: A piece of hardware including some processor and software that controls a device.

emulator: A device that mimics the behavior and functions of another device. A processor emulator.

EMI: Acronym for Electro-Magnetic Interference. This is a phenomenon by which a device or a system can generate an electromagnetic field in the radio frequency spectrum with the potential to disrupt operation of other electronic components or systems in the vicinity.

EEPROM: Acronym for Electrically Erasable and Programmable Read Only Memory.

EPROM: Acronym for Electrically Programmable Read Only Memory.

finite state machine: A device that stores the status of something at a given time, with some inputs that can change the state and/or outputs.

flag: A bit used by a program to remember something or to convey binary information to another piece of program.

flash memory: A nonvolatile memory that can be erased and reprogrammed in units of memory called blocks. The name flash memory means the memory cells can be erased in an electron tunneling process in a flash by removing an electronic charge from a floating gate associated with each memory cell.

FPGA: Acronym for Field Programmable Gate Array. A large and dense programmable logic device.

FSM: Acronym for Finite State Machine.

full-duplex: Same as Duplex.

glitch: An unwanted, transient signal transition from the current level to the other level and back to the original level.

half-duplex: Term used in communication systems that means that either the transmitter can send or the receiver can receive signals at a given time.

handshake signal: Control and feedback signals used between two (or more) devices to facilitate exchange of data.

hexadecimal: A base-16 number system. The numbers go from 0 to 9, A, B, C, D, E, and F.

host: A computer system acting as a master that provides services to other connected devices or systems.

IC: Acronym for Integrated Circuit. A semiconductor chip with many transistors and resistors connected to make an electronic circuit.

ICE: Acronym for In-Circuit Emulator. A development tool for developing microprocessor-based system and devices. The ICE mimics the operation of the target processor during the development process.

IIC or I2C: Acronym for Inter-IC communication. A communication bus with only two signals.

infrared: Part of the light spectrum that is just above that of visible light in the red end of the spectrum.

IrDA: Acronym for Infrared Data Association. IrDA is an industry-sponsored organization to design standards for the hardware and software used in infrared communication links.

instruction: Lowest level command given to the processor by a program.

interrupt: An asynchronous signal generated by a peripheral device to the processor that, when asserted, indicates to the processor to take notice and execute a special piece of program called an ISR.

I/O: Input Output. Peripheral devices of a processor to interact with the physical environment around it.

I/O map: A table containing the addresses, within the I/O space, of the input and output devices of a computer system.

I/O space: A type of addressing region that allows a processor to connect I/O devices.

instruction pointer: A special register in a processor that points to an address in the program memory from where the current instruction is being executed by the processor.

ISA: Acronym for Industry Standard Architecture. ISA is a bus architecture used in IBM personal computers. It allows connectivity between the processor and the associated peripheral circuits and devices.

ISR: Acronym for Interrupt SubRoutine. A program that is executed by the processor when an interrupt occurs in a computer system.

interrupt vector: Address of an ISR.

Kbps: Acronym for Kilobits per second. Kilo is 1000 here.

Kbyte: Kilobyte. Kilo is 1024 here.

latency: Usually associated with the interrupts in a computer system and refers to the time it takes to respond to an interrupt signal.

LED: Acronym for Light Emitting Diode. A semiconductor device that emits light when a voltage of appropriate polarity and value is applied to it.

load-store architecture: A processor architecture in which the memory is accessed using only the load and store commands. No other operations are allowed on the memory contents directly.

logic analyzer: An instrument to observe digital signals as a function of time. The logic analyzer has a fluorescent or LCD display that displays the digital signals.

logic gate: A digital circuit that has one or more inputs and an output. It performs a logical operation (such as AND, OR, XOR, XNOR, NAND, NOR, NOT) on the inputs and produces a result on the output.

microcomputer: A Microprocessor and associated support circuitry, peripheral I/O components, and memory (program as well as data) put together to form a small computer specifically for data acquisition and control applications.

microcontroller: A microcomputer on a single chip.

microprocessor: A Central Processor Unit (CPU) on a single chip.

mnemonic: An abbreviation, an aid for remembering the code of a processor.

NVRAM: Acronym for Non-Volatile Random Access Memory. RAM with battery backup for retaining the contents of the RAM when the main power is put off.

oscilloscope: An instrument to observe electrical signals as a function of time. It has a fluorescent or LCD display for observing the signals.

object code: A program that a processor can execute directly.

pipeline: Refers to an internal implementation of a processor in which the instructions are continuously being fetched by a section of the processor and placed in a queue for execution by the execution section of the processor, which in turn places the results in an output queue to be stored back to the designated destination. A nonpipelined processor, on the other hand, fetches an instruction, executes it, and stores the results before fetching the next instruction. Pipelining improves overall execution speed because of overlapping of the various stages of program execution.

PLD: Acronym for Programmable Logic Device. A digital circuit whose functionality can be changed as per the logic required. The PLD has a combination of AND, OR, and NOT gates connected through a network. By choosing the right gates and the right interconnects, the PLD can be made to implement any logic function.

program counter: Same as Instruction Pointer.

PWM: Acronym for Pulse Width Modulation. In PWM, the pulse width of the frequency is changed while keeping the frequency constant. This changes the DC value of the signal. For low width pulse, the DC value is smaller than a pulse of higher width.

RAM: Acronym for Random Access Memory. A memory device any part of which can be accessed directly. In the early days of computing, this was contrasted with tape memory, which was sequential access memory. Now RAM usually means some sort of volatile, read/write memory.

RISC: Acronym for Reduced Instruction Set Computer. A type of computer architecture with a small number of minimum instructions, characterized by a very regular instruction structure of fixed length, a load-store approach to memory access, and a large number of registers. Contrast it with CISC.

reset: To restart. A signal in a processor that initializes the internal register and control circuit to a default value and starts executing the program from the first memory location.

reset vector: Address of the reset code.

reset address: The address that the processor first accesses for the first instruction, after the reset signal is applied.

reset pointer: Address of the reset code.

reset code: The program that the user writes as part of the system initialization.

RMS: Acronym for Root Mean Square. A way to express the value of the signal by averaging the square of the signal over a full cycle of the signal and then taking the root of the average quantity. The "220-V" or "115-V" wall outlet voltage in most countries is an RMS value. For a sine wave, the peak signal is about 1.4 times the RMS value of the signal.

RS-232:   A protocol for serial asynchronous transfer of data.

simplex: Used in communication. Simplex refers to communication in only one direction and not in the reverse direction.

simulator: A software for monitoring the execution of a program. Simulator allows the user to execute the program instruction by instruction and inspect register, memory, and I/O port contents. The simulator allows sections of program to run at full speed by placing break points.

SPI: Acronym for Serial Peripheral Interconnect. A four-wire serial synchronous serial communication protocol between two devices or ICs.

stack: A read/write storage space used for storing the return address of a calling program. The stack has a Last In First Out structure. The value written last is read first.

stack pointer: An address register that points to the current top location in the stack.

startup code: A section of a program that is executed for system initialization at the very beginning.

timer: A counter that is incremented by a clock signal.

target device: Refers to the processor in the target system that is under development.

UART: Acronym for Universal Asynchronous Receiver Transmitter. A serial communication device or IC that converts a byte of data into serial bits and transmits it out at a certain rate. Similarly, it receives incoming serial bits and assembles these bits into a byte for the host.

watchdog timer: A special timer that is used specifically for the purpose of resetting the system if it overflows. A piece of program resets the watchdog timer every so often before the watchdog timer expires. However, if the program fails to reset the timer, indicating that the program has crashed or entered into some infinite loop (indicating unwanted and unexpected program or system behavior), the watchdog timer overflows and this generates a processor reset signal.

INTERNET RESOURCES FOR
THE AVR

The following Web sites provide useful information about AVR controllers, software projects, etc.

Atmel AVR Page: http://www.atmel.com/atmel/products/prod23.htm

Atmel AVR Data Sheets: http://www.atmel.com/atmel/products/prod200.htm

Atmel AVR Application Notes:
http://www.atmel.com/atmel/products/prod201.htm

Atmel AVR Support Tools:
http://www.atmel.com/atmel/products/prod202.htm

Atmel AVR Software: http://www.atmel.com/atmel/products/prod203.htm

Atmel AVR Reference Library:
http://www.atmel.com/atmel/products/prod204.htm

Atmel AVR Third Party Vendors:
http://www.atmel.com/atmel/products/prod205.htm

AVR Resource and Information Center http://www.avr-forum.com/

Omega Verksted's AVR Resource Page for projects and links
http://www.omegav.ntnu.no/avr/

Dontronics. AVR Kits, parts, projects, links, and much more. A very useful site. http://www.dontronics.com

Jack's AVR Page. Projects, code, free JAVR Basic Compiler, kits. http://www3.igalaxy.net/ jackt/ or http://www.javrbasics.com

SPJ Systems. C compiler http://www.spjsystems.com

AVR Embedded Microcontroller Resources. Links, links, and links. http://www.ipass.net/hammill/newavr.htm

More AVR links and projects. http://come.to/Stelios_Cellar

AVR in education. A microcontroller design lab using the AVR. http://instruct1.cit.cornell.edu/courses/ee476/

INDEX

1200 processor (*see* AT90S1200)
20-pin AVR, 57
2313 processor (*see* AT90S2313)
2323 processor (*see* AT90S2323)
2333 processor (*see* AT90S2333)
2343 processor (*see* AT90S2343)
28-pin AVR, 57
40-pin AVR, 58
4004 microprocessor, 2
4094 shift register, 111
4414 processor (*see* AT90S4414)
4433 processor (*see* AT90S4433)
4424 processor (*see* AT90S4434)
64-pin AVR, 58
74165 shift register, 110
8-pin AVR, 57, 58
8048 microcontroller, 2
8051 microcontroller, 2
8515 processor (*see* AT90S8515)
8574 I/O expander, 111
68HC11, 2

Access cycle:
    external SRAM, 47
    SRAM, 27, 30
Access time:
    EEPROM, 43, 267
    SRAM, 46

Access time (*Cont.*):
    external SRAM, 47
Accumulator, 13
    machine, 15
ADC instruction, 64
ADC using on-chip comparator,
    113–117
ADD instruction, 63
Address:
    Bus, 47
    EEPROM register, 37
    Port I/O, 45
    Postincrement, 27, 62
    Preincrement, 27, 62
Addressing modes:
    Data Direct, 61
    Data Indirect, 62
    Indirect Program, 62
    I/O Direct, 61
    Register Direct, 59, 61
    Relative Program, 62
ADIW instruction, 64
Alkaline batteries, 82
ALU, 2, 12, 26
Analog comparator control and
    status register (ACSR),
    42, 56, 113
Analog I/O Port, 13

Analog multiplexer, 117
Analog-to-Digital converter, 112
AND instruction, 65
ANDI instruction, 65
Application Specific IC (ASIC), 3
Arbitrary Waveform Generator,
    150–151
Architecture, AVR Processor, 22
Arithmetic and logic instructions,
    63–67
ASR Rd instruction, 61, 81
Assembler, 185–187
AstroDat: Astronomy DAS,
    261–275
    block diagram, 262
    circuit diagram, 264
    data readout, 268–270
    data storage format, 269
    design description, 263–266
    system development, 267
    user's guide, 270–275
Astronomical data acquisition
    system, 255–275
Asynchronous serial transfer, 158
AT24C512 EEPROM, 142, 263
AT90S1200/A, 57, 90, 98, 99,
    117, 124, 127, 137, 208,
    226

AT90S2313, 57, 113, 122, 126, 134, 140, 142, 148, 149, 233, 263, 301, 304, 322
AT90S2323, 57, 288, 295, 297
AT90S2333, 57
AT90S2343, 57, 91, 220, 223, 282, 284, 317
AT90S2343-based dice, 220
AT90S4414, 58
AT90S4433, 57
AT90S4434, 58
AT90S8515, 58, 106, 107, 180
AT90S8535, 192
Atmel Corporation
Autonomous data acquisition system, 257
Average value of digital signal, 124
AVR, 6
AVR Assembler, 186–187
AVR-based frequency counter, 291–300
   block diagram, 295, 296
   circuit diagram, 297
   design code, 298
   design description, 295
   testing, 300
   usage, 298
AVR controller types, 57–58
AVR family architecture, 22–25
AVR hardware design, 81
AVR instruction length, 25
AVR instruction set, 59–79
AVR interfacing, 97–155
AVR Protoboard, 309–313
   block diagram, 310
   circuit diagram, 312
   design description, 310
   ISP dongle, 314
AVR Simulator, 187
AVR Studio, 188

Backup power supply, 266
BASCOM-AVR, 198
BasicX, 198
Battery duty cycle, 82
Battery energy content, 82
Battery power, 82–83
   primary, 82
   secondary, 83
Baud rate, 158
Baud rate register, 42
Beacon, 245
Beginner's interfacing circuit, 97–99
Bit and bit-test instructions, 76–79
Bit banging UART, 108
Bit rate, 158
Branch instructions, 70–72
Brownout detector, 12

Bus:
   CAN, 182–183
   communication, 157
   IIC/I2C, 163–166
   ISA, 172–175
   Microwire, 163–164
   Universal Serial, 174–178

C compiler, 195
CALL instructions, 68
CAN: Controller Area Network, 182–183
Carry flag, 29
CBI instruction, 77
CBR instruction, 66
CCD camera controller, 318
Ceramic resonator for AVR, 87–88
CLC instruction, 77
CLH instruction, 79
CLI instruction, 78
CLN instruction, 78
Clock oscillator, 12
Clock source for AVR, 86–93
CLR Rd instruction, 60
CLS instruction, 78
CLT instruction, 79
CLV instruction, 79
CLZ instruction, 78
Code assembler, 185–187
Code authenticator, 317
Code simulator, 187, 200
COM Rd instruction, 60
Combinational Logic on AVR, 104–105
Communication links, 157–183
Comparator on chip, 113
Compare instructions, 69
Complex Instruction Set Computer (CISC) 14
Connecting PC serial port to AVR, 105–109
Constants Table Accessing, 149–150
Controller Area Network (see CAN)
Counter, 291
CPU, 2
Crystal Clock IC, 86–87

DAC Address Bits, 128
DAS: Data Acquisition System, 255
Data Acquisition System, 255–257
Data direct addressing, 61
Data Direct instruction, 61
Data direction register:
   PortB, 39
   PortD, 39
Data Indirect instruction, 62

Data transfer instructions, 72–76
DCE: Data Communication Equipment 159
DDS MICRO-C Developers Kit, 197
DEC Rd instruction, 60, 66
Decoder:
   Address
   IBM PC Keyboard, 321
Device programmer, 193
Dice, 208
Digital I/O Port, 13
Digital-to-Analog Converter, 124
Digital state machine, 2
Diode peak inverse voltage, 83
Display LED, 99, 132–135
   LCD, 135–137
   seven segment, 132
   dot matrix, 133, 136
Dongle (see Electronic lock)
Dot matrix display refresh rate, 133, 135
Driving relay with AVR, 138–140
DS1236 Dallas micromanager chip, 94
DS1233 Dallas reset generator chip, 95–96
DTE: Data Terminal Equipment, 159
Duty cycle, 124

EEPROM, 25
   byte write, 144–145
   current address read, 144, 146
   interfacing, 141–146
   page write, 144
   random read, 144, 147
   sequential read, 144
   WP pin, 142
EEPROM Address Register (EEAR), 37
EEPROM Control Register (EECR), 38
EEPROM Data Register (EEDR), 38
EEPROM in AVR, 43–45
Electronic dice, 208
   block diagram, 210
   circuit diagram, 211
   design, 211–212
   design code, 213–217
   fabrication, 217–218
   power consumption, 219–220
   powering the, 219
Electronic lock, 280
   block diagram, 279, 280, 282
   circuit diagram, 284
   design description 284–286
   fabrication, 288
Emulator, 192–193
EOR instruction, 65

Evaluation board, 188–192
Expanding I/O, 110–112
External reset flag (EXTRF), 31, 32, 248
External SRAM, 25, 47
External SRAM enable bit (SRE Bit), 30
External SRAM access wait state bit (SRW Bit), 30

Features, AVR Processor, 21, 57–58
File Register, 25
Finite State Machine, 152–154
Fractional multiplication instructions, 67
FSM state output, 153
FSM transition table, 153
Frequency Counter, 292–293

General Interrupt Mask Register (GIMSK), 29
General Interrupt Flag Register (GIFR), 29
Global interrupt enable bit, 28

Half carry flag, 29
Half duplex, 161
Hardware Lock (see Electronic lock)
Hardware Stack, 102

IAR Assembler, 187
ICALL instruction, 68
ICE: In Circuit Emulator, 192
ICE200 AVR Emulator, 192–193
ICR1H, ICR1L: Timer/Counter1 Input Capture Registers, 37
IIC Bus, 164–166
  application, 165
  Bit transfer on, 165
  Start condition, 166
  Stop condition, 166
IIC Expanders, 111
Implementing logic equation, 3
ISP: In system Programmable, 21
ISP dongle, 314
IN instruction, 76
INC Rd instruction, 60, 66
INT0 interrupt, 263
IJMP instruction, 67
Indirect Program Addressing, 62
I/O direct instruction, 61
I/O Expansion, 110–112
I/O Memory, 27–28
I/O Ports, 45
I/P Registers, 24
Input Port, 45, 110, 212

Instruction:
  Arithmetic and Logic, 63
  Bit and Bit Test, 76
  Data Transfer, 72
  Decoder, 2, 23
  Execution, 27
  pipeline, 25
  Program Control, 67
  Register, 23
  Set, 59
Interfacing:
  ADC, 112
  DAC, 124
  LCD, 135–138
  LED, 97–99
  Relay, 138–140
  RTC, 146, 148–149
  Serial EEPROM, 141–146
  Switches, 99–101, 212
Internal RC Clock Oscillator, 90–91
Internal SRAM, 24
Internal watchdog timer, 55–56
Interrupt:
  Nested, 54
  Latency, 55
  Response, 55
  Structure in AVR, 53–55
Interrupt operation, 53–55
Interrupt sense control bits, 30
Interrupt Vector, 23–24
IrDA Data Link, 178, 181–183
  FIR mode, 181
  SIR mode, 181
ISA Bus, 172–175
  Interface for AVR, 173
  Port read, 176
  Port write, 176
  Signals, 175
ISR: Interrupt SubRoutine, 302

JAVRBasic, 198
JMP instruction, 68

Keyboard Decoder, 321
Kitchen Timer, 239–243
  block diagram, 240
  circuit diagram, 241
  design description, 240–241

L297 stepper motor sequencer, 140
L298 stepper motor driver, 140
LCD signals, 136–137
Lead Acid Battery, 83
LED Display, 132–138
LFSR: Linear Feedback Shift Register, 154, 281, 305
LFSR sequence length, 155
LFSR taps, 155
Lithium battery, 82

LM335 temperature sensor, 116
Load program memory instructions, 79
Load register instructions, 72–74
Low-pass filter for PWM DAC, 124
LSL Rd instruction, 60, 76
LSR Rd instruction, 60, 76

Mains Operated Supply, 83–84
Master-slave data communication, 285
MAX110 ADC, 121–124
MAX186 ADC, 117–121
MAX186 control byte, 120
MAX186 Data Conversion, 118
MAX186 Data Readout, 118–120
MAX186 Signals, 119
MAX3100 IrDA UART, 181
MAX521 address and command byte, 131
MAX521 command bits, 132
MAX521 communication format, 130
MAX521 DAC, 126–132
MAX521 DAC Data Transfer, 127–129
MAX521 signals, 129
Mega103/603, 58
Memory Access, 27
MCU00100 development board, 189
MCU General Control Register (MCUCR), 30
MCU Status Register (MCUSR), 30, 248
Microcontroller:
  architecture, 14
  AVR, 21–22
  choosing a, 16–18
  classification, 13, 15
  components, 11–13
  developing applications with, 18–19
  market, 14
  usage, 1
Microcontroller Architecture, 14
Micropower Regulator, 85–86
Microprocessor, 2
MICROWIRE Bus, 163–164
MICROWIRE signals, 163
Minimal Instruction Set Computer (MISC) 14
Morse Code, 224
Morse Keyer, 223–231
  block diagram, 226
  circuit diagram, 227
  code, 228
  design, 225–228
  fabrication, 228
Morse Tutor, 321

Motor speed ramping, 141, 144
Motor stepper, 140
MOV instructions, 72
Multiple MAX521 in a single
    bus, 129, 133
Multiply instructions, 66–67
Musical notes, 303
Musical toy, 301–307
    block diagram, 302
    circuit diagram, 304
    design code, 305–306
    fabrication, 303

NEG Rd instruction, 66
Negative flag, 29
Nickel cadmium cell, 83
Nipper, 202
NOP instruction, 79
Nose plier, 202

OCR1AH, OCR1AL:
    Timer/Counter1 Output
    Compare Register, 36
OCR1BH, OCR1BL:
    Timer/Counter1 Output
    Compare Register, 36
On-chip UART, 58, 108
OR instructions, 65
Oscillator (see Clock source for
    AVR)
Out instruction, 76
Output port, 45

Parallel port, 166–171
    control port address, 168
    data port address, 168
    signals, 171
    Status port address, 168
PCB: Printed Circuit Board,
    200
Peak inverse voltage (PIV) for
    diode, 83
Period counter, 293–294
PLD (Programmable Logic
    Device), 3
Point-to-point communication,
    157
Pointer register, 27
POP instruction, 76
Port:
    I/O, 45
    PortB, 39
    PortD, 39
PortB data register (PORTB),
    39
PortB input pins (PINB), 39
PortD data register (PORTB),
    39
PortD input pins (PIND), 39
Power down modes in AVR,
    56–57, 247
Power from Serial Port, 84–85

Power-on reset flag (PORF), 31,
    32, 248
Power source, 81–86
Power supply pitfalls, 290
Primary batteries, 82
Program addressing
    Direct, 59–60
    Indirect, 62
    Relative, 62
Program control instructions,
    67–72
Program memory, 12
    in AVR, 23
Programmer, 193
Protoboard, 200
Prototyping techniques, 199
PUSH instruction, 76
PWM: Pulse Width Modulation,
    124
PWM DAC, 124

Quartz crystal clock for AVR,
    88–89
Quartz clock crystal for AVR, 90

R-2R ladder DAC, 124, 127
Radio Beacon, 245
    block diagram, 246
    circuit diagram, 251
    design description, 246–249
    operation flowchart, 250
RAM, 12
Random Number Generator,
    154–155
RC clock oscillator, 91
RC clock oscillator voltage
    dependence, 91
RCALL instruction, 68
Reduced Instruction Set
    Computer (RISC), 14
Register direct instruction,
    59–61
Register file, 24, 25–26
Register-memory machine, 15
Register-register machine, 15–16
Regulator:
    Micropower (see Micropower
        Regulator)
    Voltage (see Voltage
        Regulators)
Relative Program Addressing, 62
Relay Driver, 138–140
Reset, 12
Reset circuit for AVR, 93–96
Reset source, 30, 32
Resonator (see Ceramic resonator
    for AVR)
RET instruction, 68
RETI instruction, 68
RJMP instruction, 67
ROL Rd instruction, 60, 76
ROR Rd instruction, 60, 76

RS-232:
    Connector pinout, 161
    Converter, 106
    handshake lines, 160
    Levels, 106, 109
    Line driver, 106, 161
    Port, 158–160
RS-422/423, 160–161
RS-485, 161–163
RTC: Real Time Clock, 13
RZI: Return to Zero, 181

Sample and hold amplifier, 117
SBI instruction, 77
SBR instruction, 66
Secondary batteries, 83
Set instructions, 77–79
SER Rd instruction, 60
Serial ADC, 117
Serial port, 13, 105
Seven Segment Display, 132
Shift Register for I/O expansion
    110–111
Signal bounce, 100
Sign flag, 29
SimmStick, 194–195
Simulator, 187, 200
Skip instructions, 69–70
Sleep enable bit (SE), 30
SLEEP instruction, 79
Sleep mode bit (SM), 30
Smart Card, 1
SniffStick, 257–261
    block diagram, 258
    circuit diagram, 259
    data readout, 258–260
    readout port, 258
    usage, 260
Software driven UART, 108
Solder Iron, 202
SPI Bus, 163–164
SPI control register, 40
SPI I/O data register, 39
SPI status register, 39
SRAM in AVR, 46
SRAM Interface, 46–48
Stack, 101
Stack operation, 53, 101
Stack machine, 15
Stack pointer, 101
Stack pointer register (SP), 29
State machine, 152–154
STATUS register (SREG), 28–29,
    305
Stepper motor interface, 140
Steps for prototyping,
    203–205
STK200 board, 189–192
STK300 board, 192
Store instructions, 74–75
Subtract instructions, 64–65
SWAP Rd instruction, 61, 77

Swipe Card, 320
Switch-case implementation, 150–152
Switch debounce, 100–101
Switch interfacing with AVR, 99–101

Temperature logger, 318
Temperature sensor, 116, 318
Terminal emulation program, 106
Timer, 13
Timer/Counter0 control register (TCCR0), 31
Timer/Counter0 register (TCNT0), 31
Timer/Counter1 (TCNT1H/TCNT1L), 34
Timer/Counter1 control register A (TCCR1A), 32
Timer/Counter1 control register B (TCCR1B), 33
Timer/Counter1 Output compare register A (OCR1AH, OCR1AL), 35
Timer/Counter1 Output compare register B (OCR1BH, OCR1BL), 36
Timer/Counter1 Input compare register (ICR1H, ICR1L), 37

Timer0 interrupt, 228, 305
Timer1 interrupt, 263
Timer0 ISR, 305
Timer operation in AVR, 47–49
Timing diagram convention, 6–9
Tiny10, 58
Tiny12, 58
Tiny13, 58
Tiny22, 58, 91, 223, 248, 282, 284
Tiny22 for radio beacon, 249
TMS1000 Microcontroller, 2
Tools for prototyping, 202–203
TST Rd instruction, 60
Tweezers, 202
Two's complement flag, 29

UART baud rate register, 42
UART control register, 41
UART I/O data register, 40
UART operation in AVR, 49–53
UART status register, 40
ULN2003 darlington array, 138, 141
USB: Universal Serial Bus, 174–178

USB: Universal Serial Bus (Cont.):
connectivity, 177
devices, 178
host, 177
topology, 179

Voltage reference source, 117
Voltage regulators, 85–86
Voltmeter, 233
    block diagram, 234
    circuit diagram, 236

Wall plug-in transformer, 83
Watchdog control register (WDTCR), 37
Watchdog timer, 13
Watchdog timer prescale select, 38
WDR instruction, 79
Wire stripper, 202

X register, 66, 73

Y register, 66, 73

Z register, 66, 73
Zero flag, 29
Zinc chloride batteries, 82

About the Author

Dhananjay V. Gadre is a scientific officer with the Instrumentation Program of the Inter-University Centre for Astronomy and Astrophysics (IUCAA) in Pune, India. He has a M.Engr. from the University of Idaho and a M.Sc. from the University of Delhi. Dhananjay's interests include computer architecture, communication networks, hardware-software co-design, programmable logic devices and Hardware Description Languages. He is a licensed radio amateur (VU2NOX). He has written articles for a number of electronics magazines, including *Electronics World*. He is also the author of *Programming the Parallel Port: Interfacing the PC for Data Acquisition and Process Control* (R&D Books, 1998).

SOFTWARE AND INFORMATION LICENSE

The software and information on this diskette (collectively referred to as the "Product") are the property of The McGraw-Hill Companies, Inc. ("McGraw-Hill") and are protected by both United States copyright law and international copyright treaty provision. You must treat this Product just like a book, except that you may copy it into a computer to be used and you may make archival copies of the Products for the sole purpose of backing up our software and protecting your investment from loss.

By saying "just like a book," McGraw-Hill means, for example, that the Product may be used by any number of people and may be freely moved from one computer location to another, so long as there is no possibility of the Product (or any part of the Product) being used at one location or on one computer while it is being used at another. Just as a book cannot be read by two different people in two different places at the same time, neither can the Product be used by two different people in two different places at the same time (unless, of course, McGraw-Hill's rights are being violated).

McGraw-Hill reserves the right to alter or modify the contents of the Product at any time.

This agreement is effective until terminated. The Agreement will terminate automatically without notice if you fail to comply with any provisions of this Agreement. In the event of termination by reason of your breach, you will destroy or erase all copies of the Product installed on any computer system or made for backup purposes and shall expunge the Product from your data storage facilities.

LIMITED WARRANTY

McGraw-Hill warrants the physical diskette(s) enclosed herein to be free of defects in materials and workmanship for a period of sixty days from the purchase date. If McGraw-Hill receives written notification within the warranty period of defects in materials or workmanship, and such notification is determined by McGraw-Hill to be correct, McGraw-Hill will replace the defective diskette(s). Send request to:

Customer Service
McGraw-Hill
Gahanna Industrial Park
860 Taylor Station Road
Blacklick, OH 43004-9615

The entire and exclusive liability and remedy for breach of this Limited Warranty shall be limited to replacement of defective diskette(s) and shall not include or extend to any claim for or right to cover any other damages, including but not limited to, loss of profit, data, or use of the software, or special, incidental, or consequential damages or other similar claims, even if McGraw-Hill has been specifically advised as to the possibility of such damages. In no event will McGraw-Hill's liability for any damages to you or any other person ever exceed the lower of suggested list price or actual price paid for the license to use the Product, regardless of any form of the claim.

THE McGRAW-HILL COMPANIES, INC. SPECIFICALLY DISCLAIMS ALL OTHER WARRANTIES, EXPRESS OR IMPLIED, INCLUDING BUT NOT LIMITED TO, ANY IMPLIED WARRANTY OF MERCHANTABILITY OR FITNESS FOR A PARTICULAR PURPOSE. Specifically, McGraw-Hill makes no representation or warranty that the Product is fit for any particular purpose and any implied warranty of merchantability is limited to the sixty day duration of the Limited Warranty covering the physical diskette(s) only (and not the software or in-formation) and is otherwise expressly and specifically disclaimed.

This Limited Warranty gives you specific legal rights; you may have others which may vary from state to state. Some states do not allow the exclusion of incidental or consequential damages, or the limitation on how long an implied warranty lasts, so some of the above may not apply to you.

This Agreement constitutes the entire agreement between the parties relating to use of the Product. The terms of any purchase order shall have no effect on the terms of this Agreement. Failure of McGraw-Hill to insist at any time on strict compliance with this Agreement shall not constitute a waiver of any rights under this Agreement. This Agreement shall be construed and governed in accordance with the laws of New York. If any provision of this Agreement is held to be contrary to law, that provision will be enforced to the maximum extent permissible and the remaining provisions will remain in force and effect.